ENVIRONMENT AND DEVELOPMENT IN THE STRAITS OF MALACCA

The Straits of Malacca is one of the busiest and most significant stretches of water in the world. It has long provided an important artery linking economic and strategic theatres in Europe and the East. Previously exploited by Chinese and Indian traders, the era of European colonialism from the seventeenth century emphasised its geostrategic significance. Today, the Straits' location at the heart of Southeast Asian economic development has reinforced their importance.

This book represents the first in-depth study of the environment and development of the Straits, and their wider significance for the regional and global economy. The authors focus on four key areas:

- The physical environment of the Straits region.
- Analysis of past and present exploitation of the region's resources.
- The historical role of the region, for example its place in the spice trade and in colonial rivalries.
- Contemporary economic and environmental change in the region, and the challenges posed to stability by pollution and urban growth.

Mark Cleary has taught and researched in the fields of economic and historical geography in Europe, Southeast Asia and New Zealand. He is Reader in Human Geography at the University of Plymouth.

Goh Kim Chuan has established geography programmes at the University of Science Malaysia at Penang, the University of Brunei Darussalam and the National Institute of Education, Nanyang Technological University, Singapore. He is the regional editor of *Land Degradation and Development.*

ROUTLEDGE STUDIES IN DEVELOPMENT AND SOCIETY

1 SEARCHING FOR SECURITY
Women's responses to economic transformations
Edited by Isa Baud and Ines Smyth

2 THE LIFE REGION
The social and cultural ecology of sustainable development
Edited by Per Råberg

3 DAMS AS AID
Anne Usher

4 POLITICS OF DEVELOPMENT COOPERATION
NGOs, gender and partnership in Kenya
Lisa Aubrey

5 PSYCHOLOGY OF AID
A motivational perspective
Stuart Carr, Eilish McAuliffe and Malcolm MacLachlan

6 GENDER, ETHNICITY AND PLACE
Women and identity in Guyana
Linda Peake and D. Alissa Trotz

7 HOUSING AND FINANCE IN DEVELOPING COUNTRIES
Edited by Kavita Datta and Gareth Jones

8 PEASANTS AND RELIGION
A socioeconomic study of Dios Olivorio and the Palma Sola religion in the Dominican Republic
Jan Lundius and Mats Lundahl

9 ENVIRONMENT AND DEVELOPMENT IN THE STRAITS OF MALACCA
Mark Cleary and Goh Kim Chuan

ENVIRONMENT AND DEVELOPMENT IN THE STRAITS OF MALACCA

Mark Cleary and Goh Kim Chuan

London and New York

First published 2000 by Routledge
2 Park Square, Milton Park, Abingdon, Oxon, OX14 4RN

Simultaneously published in the USA and Canada
by Routledge
270 Madison Ave, New York NY 10016

Routledge is an imprint of the Taylor & Francis Group

Transferred to Digital Printing 2007

Typeset in Garamond by Steven Gardiner Ltd, Cambridge

British Library Cataloguing in Publication Data
A catalogue record for this book is available from the British Library

Library of Congress Cataloguing in Publication Data
Cleary, Mark, 1954–
Environment and development in the Straits of Malacca /
Mark Cleary and Goh Kim Chuan.
p. cm.
Includes bibliographical references and index.
1. Malacca, Strait of – Ecnomic conditions. 2. Malacca, Strait of – Environmental conditions. I. Goh, Kim Chuan, 1946– .
II. Title.
HC441.C55 2000 99-36256
333.7′09165′65 – dc21 CIP

ISBN 0-415-17243-8

Publisher's Note

The publisher has gone to great lengths to ensure the quality of this reprint but points out that some imperfections in the original may be apparent

CONTENTS

TABLES

FIGURES

PREFACE

The conception, research and writing of this book arose out of a joint interest in the role played by the Straits of Malacca in the historical and contemporary period. In seeking to understand the environmental, historical, economic and social destinies of this part of Southeast Asia, we increasingly saw the role played by the Straits as being pivotal. People, ideas, ships, trade goods, sediments, pollutants, hydrocarbons all flowed along its sea-lanes and currents, animating the life of the communities abutting its waters in both the past and present. This book was shaped by our joint conception that the Straits themselves provided a unifying theme around which we could best understand the nature and pace of environmental, economic and social change.

We owe a debt of gratitude to those individuals and institutions who have helped us in the researching and writing of this book. Our respective institutions, Nanyang Technological University and the University of Plymouth, provided valuable academic leave and financial support. The Cartography Unit at the University of Plymouth kindly drew all the figures. A number of individuals have also lent support, and we should especially like to thank Professor Chia Lin Sien and Simon Francis. We should like to dedicate this book to the Goh and Cleary families in thanks for all their support.

1

INTRODUCTION: A SEABORNE WORLD

The Straits of Malacca, today one of the busiest stretches of sea in the world, extend for some 500 miles from north to south between Malaysia and the Indonesian island of Sumatra, and its waters are shared between three states: Malaysia, Singapore and Indonesia. The Straits are narrow and crowded: at their widest they are only 126 nautical miles around the island of Penang; whilst at their narrowest, at Little Karimun, close to the adjoining Straits of Singapore, they are little more than nine nautical miles. This stretch of water, shallow, littered with sandbanks and divided up by islands large and small, together with the coastal territories, estuaries, and human settlements that abut it, has been enormously important in shaping the historical and contemporary patterns of life and livelihood in Southeast Asia. The way in which these waters, and the traffic—human, biotic and marine—upon it, have shaped this important region is the subject of this book.

That the seas, reefs, sands, islands and coasts of the Straits region have an underlying unity—that which unites them is greater than that which divides them—is the central theme which underlies the conception of the book. Around and through the Straits have flowed and coalesced many of the vitally important economic, social, political and cultural currents that have forged the character of this part of Southeast Asia, and influenced many other regions, both near and far. Along these sea-lanes came the Arab and Indian sailors and traders who met their Malay and Chinese counterparts here and gave the ports and cities of the region their distinctiveness. Gujerati merchants, traders sailing in Arab dhows, cargoes of textiles and woods came into the region in exchange for the spices and jungle products of the Peninsula. Chinese traders were equally adept at seeking both the trade goods and political loyalties of the emerging Malay and Sumatran states along the shores of the Straits.

Changes in patterns of trade and technology can also be read in the altered landscapes, cities, ports and hinterlands of these territories. The architects of European expansion—first the Portuguese, then the Dutch, later the English and French—saw the Straits as a key that would unlock the treasures of a mythical and mystical 'East'. Controlling the precious commodities of the region—so many of which had to pass through the Straits—was the prize. The spices of the Mollucas, the jungle products of the interior, the precious metals (tin, gold, silver) of the hinterlands were to bring explorers, traders, merchants, colonists and, eventually, colonisation to the region.

Whilst the importance of these external adventurers may properly, in the light of contemporary scholarship, have been lessened, the importance of that period's influences on the institutions and physical form of the region remains strong. Whilst the power and pre-eminence of some of the great cities of the Straits—Aceh, for example, Srivijaya or Malacca itself—predated European impacts by many hundreds of years, other cities—Penang, or Singapore—owed much to colonial political and economic intervention. The particular balance—perhaps symbiosis would be a more appropriate term—between the indigenous and external is displayed in fields as diverse as built form, administrative structures, business methods, religious beliefs and rituals, and systems of social organisation. It is a diversity that owes much to the openness of the seas of the region and to the role of the Straits in funneling interchange of all kinds.

To what extent can our theme of unity be grounded in the historical and contemporary character of the Straits region? The French historian, Fernand Braudel (1975, I: 276), once wrote of that great inland sea, the Mediterranean, that it:

> has no unity, but that created by the movements of men, the relationships they imply and the routes they follow . . . land routes and sea routes, routes along the rivers and routes along the coasts, an immense network of regular and causal connections, the life-giving bloodstream of the . . . region.

It is this notion of unity through movement and causal flows of goods, ideas and peoples that underlies our own conception of the Straits. Now the Straits are not the Mediterranean, nor, regrettably, do we have the skills of a Braudel. The seas here are much smaller, they are shallower and warmer and their hinterlands are much less extensive. But the role of the Straits is no less significant for that. The flows of people, goods, ideas, money, books, diseases, pollutants and ideologies have had an immense impact on the environments and peoples of the coasts and hinterlands that abut the Straits. As a seaborne world, these diverse elements share much of a common heritage, outlook and destiny. To examine them in this way is, we would argue, to remain true to their past history, their present character and their likely future.

THE STRAITS AS A UNIT

In terms of contemporary maritime law and navigation, the Straits of Malacca *per se* can be said to extend from the sandbank known as One Fathom Bank opposite Port Klang at the mouth of the Klang River, through to Tanjong Piai and the island of Little Karimun at which point ships enter into the Singapore Strait, usually divided into the Durian Strait, the Phillip Channel and Singapore Strait proper. In strict navigational terms the Straits comprise those seas between a line to the northwest:

> . . . from Ujung Baka (5.40′N, 95.26′E), the northwest extremity of Sumatra to Laem Phra Chao (7.45′N, 98.18′E), the South extremity of Ko. Phukit, Thailand.

Figure 1.1 The Straits of Malacca in their regional context.

and to the Southeast:

> . . . from Tanjung Piai (1.16′N, 103.31′E), the south extremity of Malaysia, to Pulau Iyu Kecil (1.11′N, 103.21′E), thence to Pulau Karimum Kecil (1.10′N, 103.23′E), thence to Tanjung Kebadu (1.106′N, 102.59′E).
>
> (Hamzah, 1997, 1)

Figure 1.1 shows the geographical situation of the Straits within Southeast Asia and Figure 1.2 the basic boundary units of the Straits themselves. As we shall see in a later chapter, issues of territorial sea boundaries, traffic separation schemes and marine safety requirements are complex and politically sensitive.

The Straits are essentially funnel-shaped, opening out to the Andaman Sea and Indian Ocean to the north and tapering to the southeast, through the Singapore Straits, before opening into the South China Sea. At the wide western entrance, the depths range from 34 to 84 m, but depths diminish to only 18 or 19 m close to the Aruah Islands. Further south, off Port Kelang, are a series of shoals and sandbanks, including One Fathom Bank where depths can be as low as 10 m. Depths increase again into the Singapore Straits though here the navigable channels are narrow (Chia and MacAndrews, 1981, 252–254).

For the purposes of this book we adopt a broader view of the Straits which reflects the historical rather than contemporary nature and divisions of the region. The Straits take their name from the ancient city of Malacca (Melaka in the Malay), founded in the early fourteenth century, but which can be seen as one and the latest in a line of

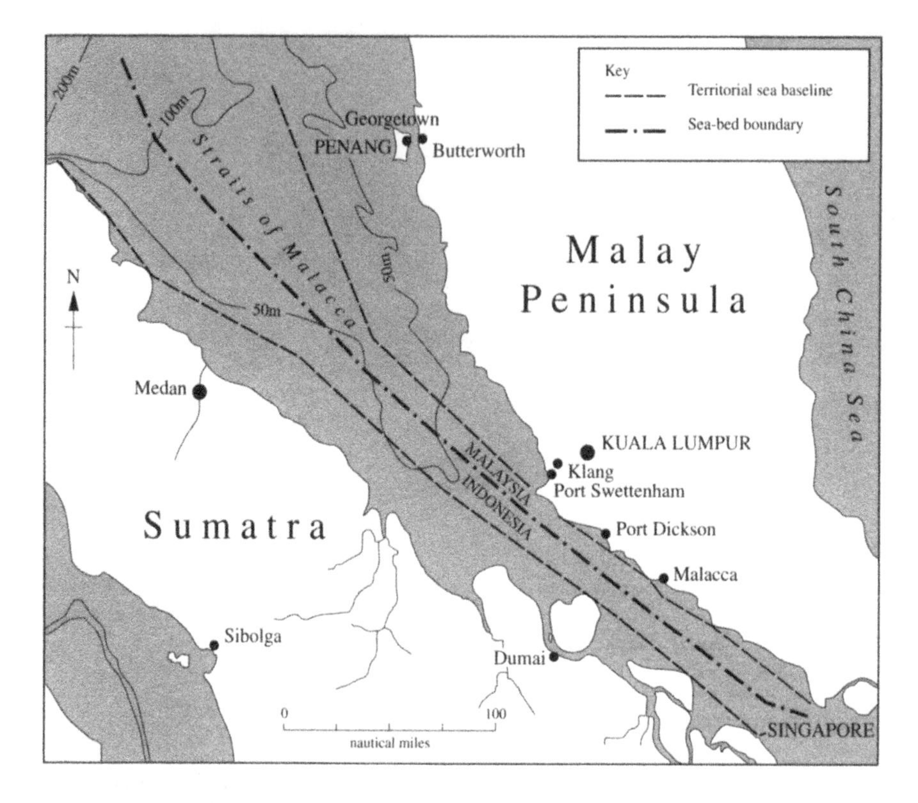

Figure 1.2 Boundaries in the Straits of Malacca.

coastal settlements fronting the Straits which developed along the trading routes between India and China. For probably two thousand years, ships and traders from India and China had been travelling the Straits—buying and selling goods which could be traded on to more distant markets. Those ports and cities that developed along the Straits thus shared common economic characteristics and shared too an openness to outside religious, cultural and political ideas. Our concept of the Straits is thus much wider than a purely navigational one, although clearly navigation is at its heart. We argue that the coast and hinterland of northern Sumatra and the key city of Acheh (henceforth Aceh), together with the east coast of Sumatra down to Palembang as well as coastal southeast Sumatra have fallen, and continue to fall, within the cultural, political and economic orbit of the Straits. To the east of the Straits, Penang island and Perak, coastal Selangor and Port Klang, the port of Kuala Lumpur, together with Melaka (henceforth Malacca), Johor and Singapore can be seen as part of a cultural region which extends well beyond the narrow confines of the Straits themselves but which was intimately shaped by the flows and movements along that waterway. In terms of contemporary administrative boundaries, our region comprises Aceh, North Sumatra, Riau and Jambi in Indonesia, Perlis, Kedah, Perak,

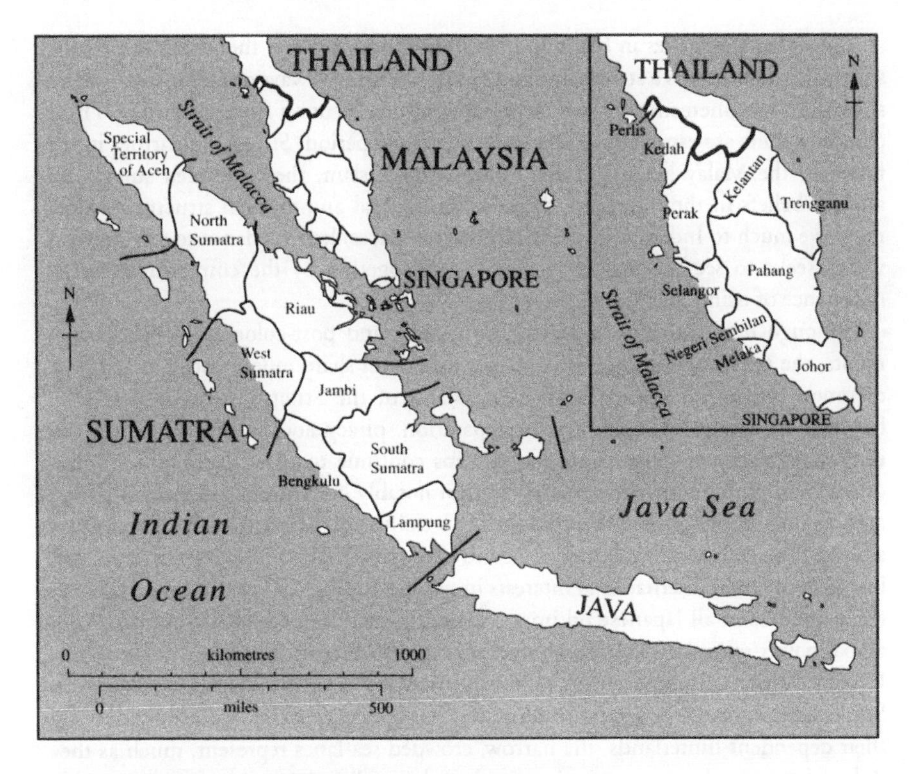

Figure 1.3 Administrative divisions in the Straits region.

Selangor, Negeri Sembilan, Malacca and Johor in Malaysia and the Republic of Singapore (Figure 1.3).

The elements that make up this culture region are varied. In terms of environment and ecology the lands around the Straits share many similarities. The tropical climate with a relatively short monsoon period, a low-lying, mangrove-fringed coast (especially on the Sumatran side) and, most importantly, a regime of winds which has played a vital part in determining the nature and rhythms of navigation through the Straits, are amongst features shared by the region. Relatively shallow, warm coastal waters have provided a range of fishing and collecting opportunities to local peoples. The coastal area itself has proved difficult to settle. It is low lying and susceptible to considerable variations in low-tide levels, which has accentuated the importance of riverine locations as the prime sites for the port towns and cities which have given the region its character. Relations between the coast and interior have always been important. In both Sumatra and Malaya, significant indigenous groups in the interior such as the Batak or *orang asli* have long traded with coastal peoples. The juxtaposition of these two ecological and cultural zones no doubt further underpinned the attractiveness of the Straits to foreign traders and merchants.

Given the similarities in habit and habitat on both sides of the water, it is hardly surprising that there is a strong unity of cultural traits and ways of life in areas which, since the early nineteenth century have been politically cut off from each other. Those common traits are especially evident in the historic period. Strong similarities in the forms of the Malay language, the dominance of Islam, the continued survival of animist beliefs in the interior and traditions of legal and cultural structures which may owe much to Indian (especially Hindu influences) lend credence to an approach which seeks to see the region as one, drawn together by the common historical experience of using the Straits as a means of livelihood.

Although now formally separated by colonial and post-colonial political settlements, the territories on each side of the Straits still share much. Most striking is a common interest in the use and management of the Straits. Issues such as legal boundaries, navigation rights, traffic separation, piracy and pollution are pressing international issues. Huge tonnages of ships continue to pass through the Straits, drawing on both international markets (most notably in hydrocarbons), and on local and regional feeder ports. At the heart of this traffic, the city state of Singapore plays a pivotal role in both stimulating and controlling sea-traffic through the Straits. And, just as in the past, international interests in passage through the Straits are vital. Over three-quarters of all Japanese oil imports pass through the Straits, giving that major global power enormous vested interests in a range of maritime issues in the region. Equally, as the economic power of Southeast Asia has grown, passage through the Straits has assumed ever greater importance. For the ports of the region, as well as for their dependent hinterlands, the narrow, crowded sea-lanes represent, much as they did in the past, a key to economic growth and to political co-existence, cooperation and stability.

THE STRUCTURE OF THE BOOK

The approach taken by the book has determined its structure, and the treatment is both thematic and chronological. Part 1, *The shaping of the environment*, seeks to outline the important geological, morphological and ecological characteristics of the region. It is our contention that without an appreciation of these elements it is difficult to sustain the arguments made about the essential unity of the region. An understanding of the nature of the sea-bed, the characteristics of coastal erosion and accretion, the ecological character of the coastal zone, and the physical structures of coast and hinterland is vital if we are to understand the myriad ways in which human groups—both indigenous and foreign—have exploited, managed or altered those environments to their own ends. To this end, Chapter 2 examines the geological, oceanographic and geomorphological character of the Straits region, and Chapter 3 considers the key hydrological, ecological and climatic features that have impacted on human settlement and growth in the area.

The relationship between the physical and cultural environments of the Straits has always been a highly dynamic one. Techniques of exploitation and management change from one society and one time period to another, and those changes can give important insights into how the environments of the region were perceived, evaluated and utilised. These changing resources and techniques are the subject of Part 2, *Resources and techniques.* In Chapter 4, we examine the range of resources of the region—soil, biotic and aquatic—and consider how use of those resources has transformed settlement and economy in the region. Important issues of resource management are also raised here as a prelude to a more detailed examination in a later chapter. In Chapter 5, some of the key relationships between technology and the use of the Straits are considered. In a region such as this, long wedded to international maritime trade, changing shipping technology has been crucial. The arrival of technologically advanced European vessels in the fifteenth, the invention of the steam ship in the nineteenth, or the development of new techniques of containerisation or bulk transport in the twentieth century, all wrought important changes in the human geography of the Straits region.

In Part 3, *Collective histories,* these varied threads—environment, technology, human settlement and use of resources and political evolution—are drawn together in a chronological examination of the historical geography of the Straits region. The central importance of maritime trade underpins each of these chapters. Chapter 6 examines the important pre-European polities in the region from Srivijaya and Malacca to Aceh and Johor, and considers the role of both short- and long-distance trade in shaping the character of these important settlements. Chapter 7 focuses on the wider impact of European intervention in the Straits from the arrival of the Portuguese off the city of Malacca in 1511 to the rise of Singapore as a global port by the late 1920s.

Part 4, *Collective opportunities,* examines the likely impacts of contemporary development on the varied human and physical environments of the Straits. Chapter 8 considers the patterns of political and economic development since the end of the Second World War, a period marked by high growth rates, the increasing impact of environmental change, and the growth of trade in both goods (from hydrocarbons to refrigerators) and people (from legal and illegal migration to the impact of international tourism). As economic growth, mass tourism and urban redevelopment reshape the region, Chapter 9 considers the likely effects of increased economic cooperation between the countries of the Straits on its future. As ASEAN (the Association of South East Asian Nations) seeks to increase cross-border cooperation amongst its members, is greater cooperation in environmental and economic matters going to result in changes to the management and future development of the region? Chapter 10 examines this issue further in relation to a range of important environmental issues that affect the Straits such as waste management, pollution, toxic waste and the law of the sea.

What we are seeking to do in this book is to provide a context within which to understand the shape, character and evolution of a distinctive and important region of the world. Our treatment is not novel and, as will soon be apparent, we have relied heavily on a number of important and influential writings in developing the content of the book. We should make clear at once the debt we owe to a number of works on the region. Sandhu and Wheatley's (1983) *Melaka*, a two-volume, edited work, provided an important source for materials relating to the city, past and present. Reid's (1988, 1993) two-volume, *Southeast Asia in the Age of Commerce, 1450–1680*, was equally full of rich insight and information. For the contemporary period, the excellent *Straits of Malacca: international co-operation in trade, funding and navigational safety* edited by Hamzah (1997), provided an important source reference to a range of often complex issues.

Part 1

THE SHAPING OF THE ENVIRONMENT

2

PHYSICAL STRUCTURES

The importance of the Straits of Malacca as an artery for human development has been noted in the introduction; the purpose of this and the following chapter is to examine the nature of the physical environment of the Straits before considering their impact on the human environment. The geology, climate, soils and vegetation of the Straits region have played an important part in shaping the ways in which different groups have evaluated and used the resources of the region. An understanding of the physical characteristics of the Straits can, we hope, provide a better appreciation of the extent of this influence on the human ecology and the economy of the littoral states, as well as the economic wellbeing of those countries whose ships ply these waters.

THE GEOLOGICAL SETTING

Southeast Asia occupies a small but geologically highly fragmented part of the earth's surface. Not only can the region be divided into the older continental part comprising Burma, Thailand, Laos, Kampuchea and Vietnam, and the insular and archipelagic part comprising the rest of Southeast Asia, this latter geographic sub-region has been splintered into thousands of islands of varying sizes. This characteristic geological fragmentation of the region has long been an enigma and in the past has given rise to considerable speculation as to origins and evolution; more scientific explanation came rather later. By the turn of the twentieth century there was sufficient information on the geology and mineral resources of the region, gathered largely by the respective colonial governments, to begin to develop a range of models of the geological evolution of the region as a whole.

In spite of the academic advances in this period, it was primarily through large scale exploration for oil and gas, and through international research efforts such as those supported by the Committee for the Co-ordination of Joint Prospecting for Mineral Resources in Asia Offshore Areas (CCOP), and the Intergovernmental Oceanographic Commission of UNESCO (IOC), that a wealth of geologic and paleogeological data was obtained. The combined research efforts of both UNESCO and CCOP was recognised in the publications of the Studies in East Asian Tectonics and Resources (SETAR). It has been largely these collaborative research efforts that have made possible the correlation of previously known data on land with the submarine topography and its tectonic elements (Hutchison, 1996).

Advances in the understanding of plate movements and global tectonics have also contributed towards a much more accurate reconstruction of the landscapes of the Straits region. Some areas of uncertainty still exist, however, and therefore the models presented by scholars are by no means conclusive. Nevertheless, based on what is available and on our current knowledge of geophysics and tectonics, a more informed understanding of the evolution of the whole of the region can be gained. Consequently, a proper appreciation of the evolution of the Straits of Malacca, which constitutes an integral part of this geological mosaic, has been developed.

The area of the Straits is found within a region characterised by complex and multiform plate movements. Several plates can be identified including the Eurasian, Indo-Australian and Pacific plates. The region thus lies within the classic convergent margin of these plate interactions, characterised by the subduction of the Indo-Australian oceanic lithosphere beneath the Eurasian continental plate, a subduction which occurs directly under Sumatra (Hanus *et al*, 1996) at an estimated rate of 67 mm per year (Demets *et al*, 1990). In this process of convergence of the horizontally moving continental Asian plate with that of the Indo-Australian plate, the higher density oceanic plate subducts and under-rides the continental plate. At this margin of active subduction are found the long submarine trenches of the outer island arcs of Indonesia. It is this subduction of the oceanic plate that accounts for the seismically and volcanically unstable island of Sumatra, which has had implications for the evolution of the Straits of Malacca.

It is clear from plate tectonics theory that the plates themselves are not of recent origin. In fact within Southeast Asia as a whole, five distinct tecnostratigraphic terranes can be recognised which accreted to each other in the Palaeozoic and Mesozoic. Most of these terranes were derived from the margins of the original Gondwanaland, initially located close to the South Pole. Much of what is within the Straits of Malacca and the western part of Peninsular Malaysia, together with the eastern part of Sumatra, fall within the Sibumasu (Sino-Burma, Malaya and Sumatra) terrane. This Sibumasu terrane is characterised by Late Carboniferous–Early Permian glaciomarine diamictites, cool-water faunas and paleontological affinities to the Gondwanaland of northwest Australia. In the case of Sumatra, for example, occurrences of the pre-Mesozoic granites and other intrusive bodies of crystalline schists, as well as Carboniferеous sedimentary units (van Bemmelen, 1949; Hamilton, 1979), suggest that part of the Sumatran crust predates the opening of the Indian Ocean, and is thus Gondwanaland in its affinities (Gasparon & Varne, 1995).

From the plate reconstruction of Southeast Asia discussed above, it is clear that collision tectonics has played an important role not only in shaping the present-day configuration of this area (Lee & Lawver, 1995), but also in accounting for much of the seismic and volcanic activity in the region. The region is seismically very active and contains many of the world's most active volcanoes. The region is also of interest in providing evidence of the relationships between tectonic evolution and economic mineral and hydrocarbon deposits (Hutchison, 1996).

Whilst deep earthquakes occur only infrequently and at the interface of sliding plates, on the landward part of Sumatra a dextral strike-slip fault, the great Sumatra Fault, is a major source of numerous earthquakes (Katili & Hehuwat, 1967). The Sumatra Fault is more than 1500 km long and runs through the entire length of the island, coinciding with the Barisan Mountain Range. Most earthquakes in Sumatra are of shallow focus associated with 11 linear fracture zones (Hanus *et al*, 1996) and these have foci of around 100 km in depth. The few deep earthquakes occur mainly in South Sumatra close to Java where most earthquakes are deep focus at locations of more than 600 km deep. There is a range of volcanic activity associated with these faults and earthquakes and Figure 2.1 shows the distribution of volcanic activity in Sumatra (Katili, 1985).

This preponderance of shallow earthquakes in Sumatra relative to Java can be explained by the late Cenozoic clockwise rotation of about 20 degrees about an axis located in or near the Sunda Strait, between 10 and 4 million years BP (Ninkovich, 1976). In the process of rotation, Sumatra and Peninsular Malaysia were pushed northeastwards for about 500 km along the system of presently inactive faults. Only after the rotation ceased did under-thrusting of North Sumatra begin, producing a short and shallow Benioff zone. The subducted slab of oceanic lithosphere is short and shallow under Sumatra, especially in the middle and northern parts. Nevertheless, the absence of a deeply penetrating subducted slab makes the over-riding and subducting plates strongly coupled in Sumatra, and powerful earthquakes can thus occur.

Given this convergence of three plates, one would expect no area within this region to be stable. Yet the Straits of Malacca and indeed their adjacent landward parts of Peninsular Malaysia and Singapore are areas of relative stability. This is because much of Sumatra east of the great fault line lies within the stable aseismic Sundaland together with Peninsular Malaysia and Borneo. In fact by the early Miocene (20 ma) the position of Peninsular Malaysia was very much as it is today (Lee & Lawver, 1995), although the North Sumatra basin was still undergoing extension (McCabe *et al*, 1988). Understandably, because of its closeness to the subduction zone, eastern Sumatra not only experiences frequent tremours and earthquakes, but the landscape itself is dissected by large faults and contains a number of volcanoes, both active and extinct.

A classic example of a massive relict feature of past eruptions is the Toba Caldera, with dimensions of 100 × 30 km. Based on K-Ar (potassium-argon) dating of core samples from the Indian Ocean and Peninsular Malaysia, Ninkovich *et al* (1978) placed the eruption in the Late Pleistocene, some 75,000 years ago. However, what is impressive is the extent of tuff deposited in an area which occupied some 20,000 to 30,000 sq km, with a thickness of several hundred metres. Away in Peninsular Malaysia, at a site in Kota Tampan, volcanic ash up to 3 m thick was first reported by Scrivenor (1931), and suggested by van Bemmelen (1949) to have originated from that eruption. One of the greatest volcanic eruptions in modern history was the eruption in 1884 at Krakatoa in South Sumatra, which is situated farther away from the Straits of Malacca, close to the Sunda Strait.

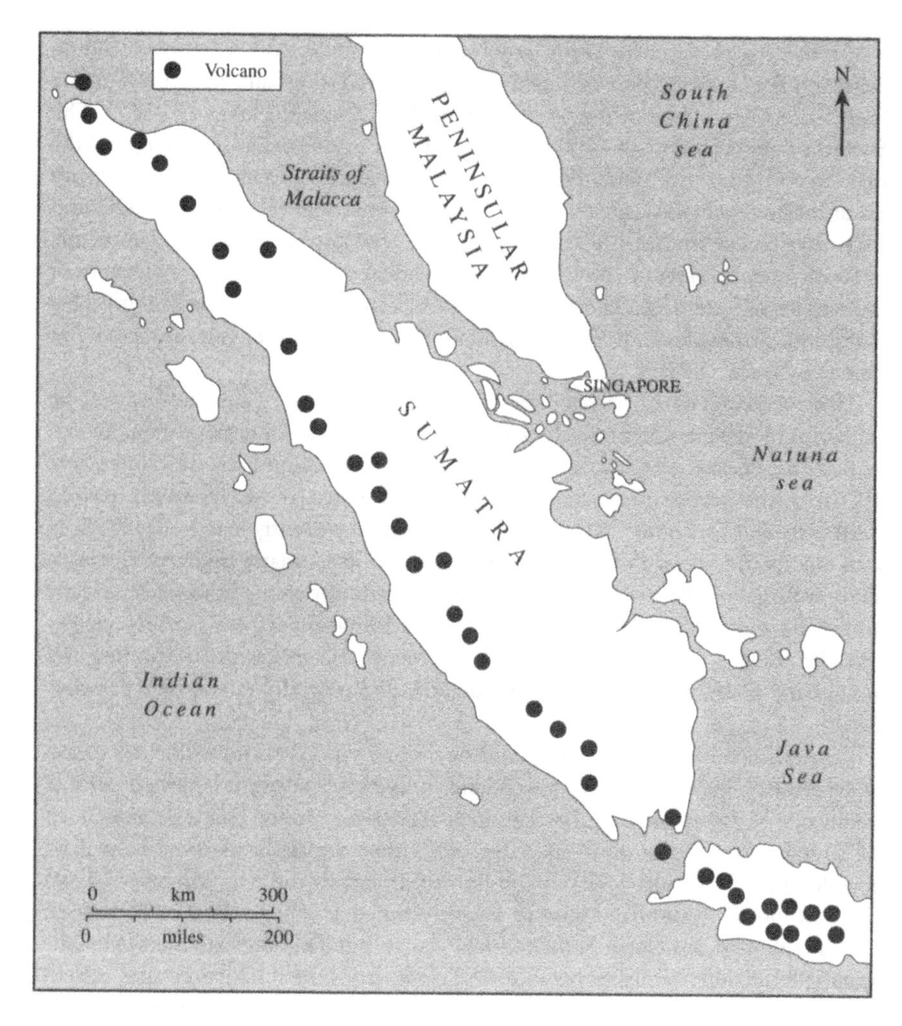

Figure 2.1 Distribution of volcanoes in Sumatra and North Java (*source*: Tomascik *et al*, 1997, 34).

Nevertheless, if the above two cases are any indication of what might happen, the implications of such eruptions for the Straits of Malacca and the rest of Southeast Asia would be devastating. Extensive blanketing of the atmosphere by ash and significant deposition of such materials of several metres thick on both land and sea would have serious repercussions on the environment of the Straits. Certainly the depth of the water body would be significantly reduced.

Whilst it is clear that Sumatra is an area of tectonic instability with frequent earthquakes—the last serious one affecting Jambi in 1996 with the epicentre in western Sumatra—the contemporary effects of such earthquakes are not especially serious as far as the Straits of Malacca are concerned. As the eastern part of Sumatra is relatively stable and the Straits themselves are relatively narrow with limited fetch, there is little danger of earthquakes off-shore producing destructive tidal waves or tsunamis which might endanger shipping in the waterway or threaten coastal cities and settlements.

What is perhaps important to note is the potential impact of this seismic activity on human settlement. The last two decades have been periods of major construction activity in Singapore and many of the coastal towns and cities abutting the Straits. High-rise buildings constructed near the coast and in areas underlain by soft soils are a cause for concern. Drawing lessons from the 19 September 1985 Mexico earthquakes, Sun and Pan (1995) inferred the potential dangers and damages to buildings in Peninsular Malaysia and Singapore from distant earthquakes located some 300 km away in Sumatra. Consequently, in both Peninsular Malaysia and Singapore a network of seismometers has been set up to monitor crustal movement. Research on building designs which would enable high-rise buildings to withstand tremors has also been given greater consideration. Obviously the problems of earthquake damage to tall buildings would be more serious for towns on the Sumatra side, even in the eastern more stable parts. Towns such as Medan and Banda Acheh have seen major growth in the last two decades and construction activity has greatly increased. The potential risk to high-rise buildings from seismic activity is not inconsiderable.

THE IMPACT OF QUATERNARY SEA-LEVEL CHANGE

From the point of view of tectonics, the land on the western side of the Strait of Malacca is certainly dynamic and has exerted an important influence on the region. But it has been fluctuations in the waters themselves that have had the most far reaching influence on the adjacent lands bordering them. In geological terms the Straits have only very recently been filled with water. Studies of the quaternary environments of the Straits show that much of the waterway within the Sundaland complex was subaerial and continental between 40,000 and 10,000 BP (Geyh *et al*, 1979). It follows, then, that there has been a sustained period of sea-level change to bring the Straits to their present state. Many studies have been carried out to chart the rise of sea-level within the region and the Straits of Malacca, studies which have also sought to establish causative factors, particularly regarding the relative roles of eustatic and isostatic adjustments (Geyh *et al*, 1979; Keller & Richards, 1967; Kudrass & Schluter, 1994).

It is not possible to fully understand the nature of coastal development in the Straits without reference to the suite of sea-level changes in the past 20,000 years to 30,000 years BP. A range of theories about the timing and extent of the rise and fall of sea-level within Southeast Asia and the Straits of Malacca have been put forward. Some consider that the sea-level at about 30,000 BP was comparable with that of the present, while others assume a considerably lower sea-level at that time (Morner, 1971). There is also the general view that sea-level underwent a steady rise from about

15,000 BP until 6,000 BP and that since then the sea-level has remained relatively unchanged. This theory, however, has been disputed as new data has suggested that the changes were far more complex than these general models might suggest.

The most likely scenario is that there was a steady rise in sea-level from about 18,000 years BP, and this continued more rapidly from 15,000 to 10,000 BP (10 m per 1,000 years). It is also likely that many rivers could not fill their valleys in that period because sedimentation could not keep pace with the rise in sea-level. This could perhaps explain the many drowned river valleys found in most of the Sunda shelf areas, including the relict delta of the northern part of the Strait of Malacca (Emmel & Curray, 1982). The sea-level curve suggests that the rise in sea-level tailed off after 8,000 BP to 2–4 m per 1,000 years. Because of the slow rate of sea-level rise, sedimentation probably played a more important role during this period. A rising sea-level also created very favourable circumstances for mangrove vegetation growth and consequently extensive swamps developed along the coast. Sea-level rise stopped at about 6,000 BP, although, here again, some have put a different date of 5,000 BP when the sea level was 2–3 m above present level. That period of higher sea-level was to last some 2,500 years. Figure 2.2 indicates the approximate land and sea boundaries in the region during the Pleistocene.

In summary, the history of quaternary sea-level rise and fall in the Straits of Malacca is very much influenced by the eustatic effects of the glacial and interglacial periods. Sea-level was at its lowest at the last glacial maximum of about 18,000 years ago. During this glacial maximum, the Straits were subaerial, as was much of the Sunda platform linking Sumatra, Java, Borneo and much of the South China Sea and the Gulf of Thailand. While 120 m below present was widely recognised as the depth of the shoreline then, others have put it at values of 150–165 m below present off the coast of north Australia, and at 146 m below present in a study of the Straits of Malacca (Emmel & Curray, 1982). Since the last glacial, which began about 14,000–18,000 years BP, melt-water has dramatically increased sea-levels at a rapid rate of 30 mm per year. This postglacial ice melt terminated around 6,000 years ago, and for much of Southeast Asia this level has remained much the same as the present level. However, there is evidence to suggest that sea-levels rose to between 2–3 m above present levels in mid-Holocene (between 2,500–6,000 years BP).

In the Straits of Malacca evidence of higher sea-levels in the Holocene comes from several sources. One important source of evidence is the occurrence of many raised beaches along both coasts. Along the west coast of Peninsular Malaysia, raised beaches in the Langkawi islands have been described by Scrivenor and Wilbourne (1923) and Jones (1981), and coastal Kedah and Perlis by Bradford (1972) and Jones (1981). Raised beaches along the west coast of Peninsular Malaysia are also described by Teh (1989). Raised beaches, too, occur in Pulau Bidan directly opposite the Merbok estuary. Bosch (1989) showed the widespread occurrences of raised beaches on both sides of the peninsula in the Quaternary Map of Peninsular Malaysia.

Because of the direct influence of waves, most raised beaches are usually aligned parallel to the present shoreline, but they can be distributed from close to the present shoreline to up to about 8 km inland in the Malay Peninsula. Teh (1989) showed

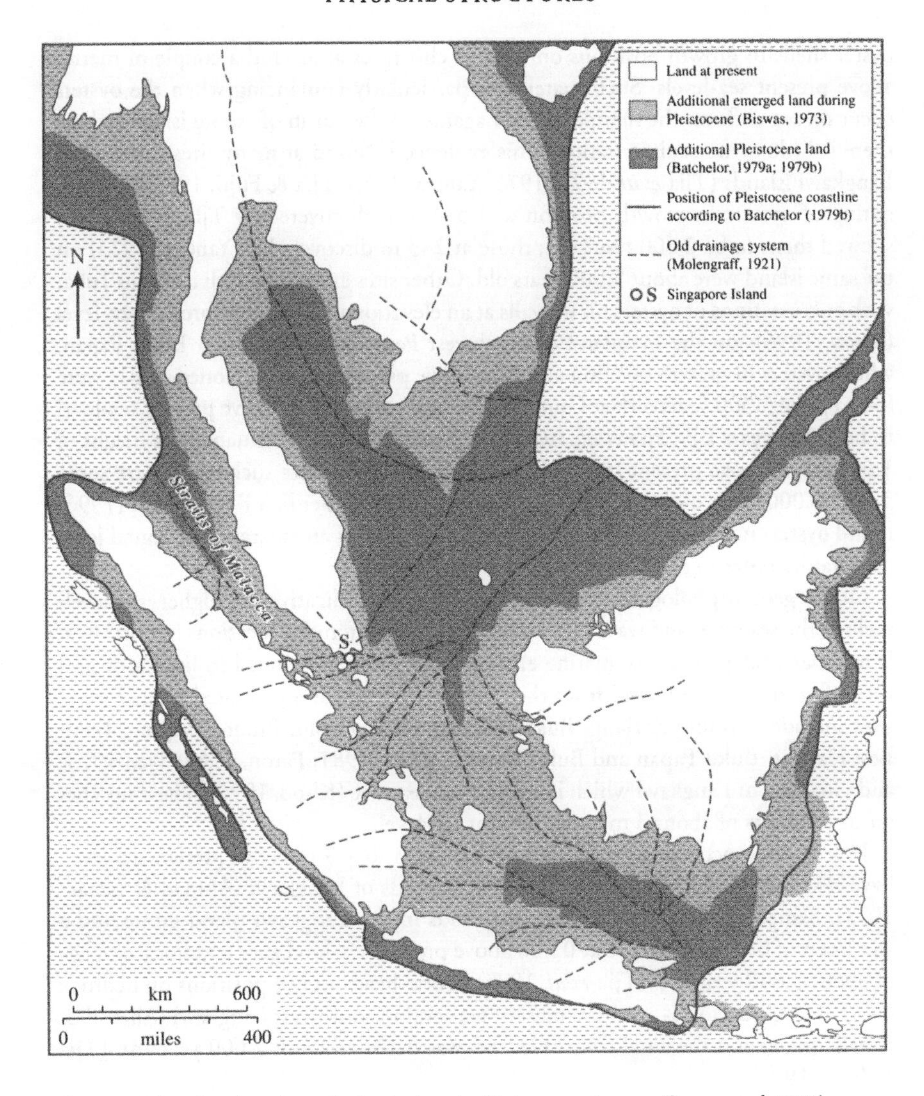

Figure 2.2 Land and sea boundaries during the Pleistocene (*source*: Gupta *et al*, 1987).

that these are composed of sand and can attain a thickness of up to 8 m, with ridge crests at <9, 9–11 and >15 m. There is some dispute about the 15 m beach ridges, however, based on the different materials found at this level which might arguably be colluvium. The age of formation of beach ridges is estimated, based mainly on radiocarbon dating of shells and plant materials, to be Holocene (developed between 2,500–6,000 years BP) except for some beach ridges inland which could be Pleistocene relics.

A second source of evidence for higher sea-levels in the Holocene is the presence of oyster shells in growth positions on rock or cliff faces at around a couple of metres above present sea-levels. Such material is particularly convincing when the oysters occur on sea cliffs on the mainland right against the sea or in off-shore islands where there is little or no sedimentation. This evidence is found at many sites such as the Langkawi islands (Tjia *et al*, 1972; 1975; Yancey, 1973; Tjia & Fujii, 1989). Carbon dating of oysters in growth position at 1.5–1.8 m discovered by Tjia *et al* (1972) showed them to be 2,600 years old; those at 2–3 m discovered by Yancey (1973) on the same island were about 5,000 years old. Other sites are at Gua Keli in Pulau Tuba, with encrustations of oysters on its walls at an elevation of 3 m above present sea-level (Khoo, 1996); on the mainland of northwest Peninsula Malaysia in Bukit Papan, where oysters in growth position at 2.8 m above present sea-level (Jones, 1981) were found to be 5,300 years old; at Gunung Keriang, about 1.5 m above present sea-level to be 4,200 years BP (Tjia *et al*, 1975); on sea cliffs at Bukit Keluang, at a height of 1.65–3.0 m above present sea-level (Tjia *et al*, 1975), where such oysters are some 3,000–6,000 years old. Off Bukit Keluang, at Pulau Perhentian Besar, Khoo (1996) found oysters in growth position on the under surface of an enlarged horizontal joint of a granite outcrop at 1.5 m above present sea-level.

Other geomorphological features that may be indicative of higher sea-levels include the sea caves and wave cut notches that occur in many locations in northwest Peninsular Malaysia. These marine erosion features are developed in limestone and they often occur well inland from the present shoreline at an elevated level. Typical sites include Gunung Keriang, Bukit Kalong, Bukit Kaplu, Bukit Chuping, Bukit Besi Hangat, Bukit Papan and Bukit Kepala (Jones, 1981; Paton, 1964; Tjia, 1973) and Gua Keli in Langkawi which is an elevated sea cave (Khoo, 1996). These notches are at positions of about 3 m above present sea-level.

Beach rocks, occurring at elevated positions up to 1.3 m above present sea-level, have also been reported from the southwest islands of Langkawi (Ahmad & Ismail, 1972). Samples of beach rock and coral heads in the Langkawi island group and a specimen of beach rock there at 0.7 m above present sea-level gave a radiocarbon age of about 3,600 years BP (Tjia *et al*, 1975). Coral heads in live positions on bedrock occur at levels up to 2.4 m above present sea-level in southwest Langkawi (Ahmad & Ismail, 1972) and the coral from this level was dated at about 2,600 years BP (Tjia & Fujii, 1989)

This range of evidence strongly suggests that the sea-level during the past several thousand years was higher than present in the Thai–Malay Peninsula, on both the east and west coasts. The height was probably less than 6 m, but generally at least 2–3 m higher than the present. It is also evident that higher or lower sea-levels of even a metre or two would have had a significant influence on coastal processes and landforms, and, within the present context, on land use and settlements along or near the coast. The Straits of Malacca were thus wholly exposed when the sea-levels were very low during the glacial maximum about 18,000 to 20,000 years BP, and submergence of a larger area than the present during mid-Holocene (between 2,500 to 6,000 years BP). It is apparent, then, that the current sea-level has probably been maintained for the past 3,000 years.

What, then, were the Straits of Malacca like when they were part of the exposed Sundaland during the glacial maximum? The character of the Straits before submergence can be reconstructed through evidence gained from bathymetry, sea-floor echo characteristics and shallow sub-bottom information. This was the approach taken by Emmel and Curray (1982), who discovered the existence of a complex Pleistocene lowered sea-level alluvial-delta-fan system in the main Straits. They showed four major environments that constituted the Malacca Strait. These were:

1 The narrow shoal part, south of 3° 30′N showing a rugged sub-bottom at shallow depth of pre-Pleistocene topography. Here the area is capped by acoustically transparent deposits, which in turn are overlain by folded sediments. Sand waves are abundant in this region.
2 A shallow shelf off Sumatra and the Malay Peninsula, lying adjacent to the delta but generally less than 60 m water depth. This was the upper flood plain environment where major river channels and smaller cut and filled valleys were numerous. There is also evidence of filling of shallow basins and depressions.
3 The submerged delta, located in the northwest approaches to the Malacca Strait. This prograded region is situated seaward of a broad Pleistocene valley, through which the confluent Sumatra rivers and rivers from Malaysia south of 4° drained during periods of low sea-level.
4 The northwestern slope leading to the Mergui–North Sumatra basin, with, in the northeastern part, a small fan developed at a depth of 300 and 400 m.

In essence, the main part of the Straits was deltaic in form, situated just beyond the boundaries of a broad valley, in which the major rivers of Sumatra and Peninsular Malaysia were confluent during lowered sea-levels (Emmel & Curray, 1982, 215). The geomorphological processes taking place during this early period of very low sea-level must certainly have influenced much of present-day bathymetric characteristics of the Straits of Malacca. The more recent period of higher stand-still starting from 2,700 years BP to about 650 years BP, and the present sea-level starting some 600 years ago would have influenced much of the coastal processes and development of the west coast of Peninsular Malaysia and the east coast of Sumatra.

PHYSIOGRAPHIC CHARACTERISTICS OF THE STRAITS

The 'physiographic catchment' of the Straits region is defined by the Barisan Range in western Sumatra in the extreme west, and the Main Range of Peninsular Malaysia in the extreme east. The almost parallel northeast–southwest alignment of both mountain ranges which run the whole length of the Straits of Malacca (and indeed almost the whole length of the respective lands) makes the study of the physical characteristics of this 'physiographically natural basin or unit' relatively neat. A better appreciation of the submarine character of the Straits of Malacca can be gained from

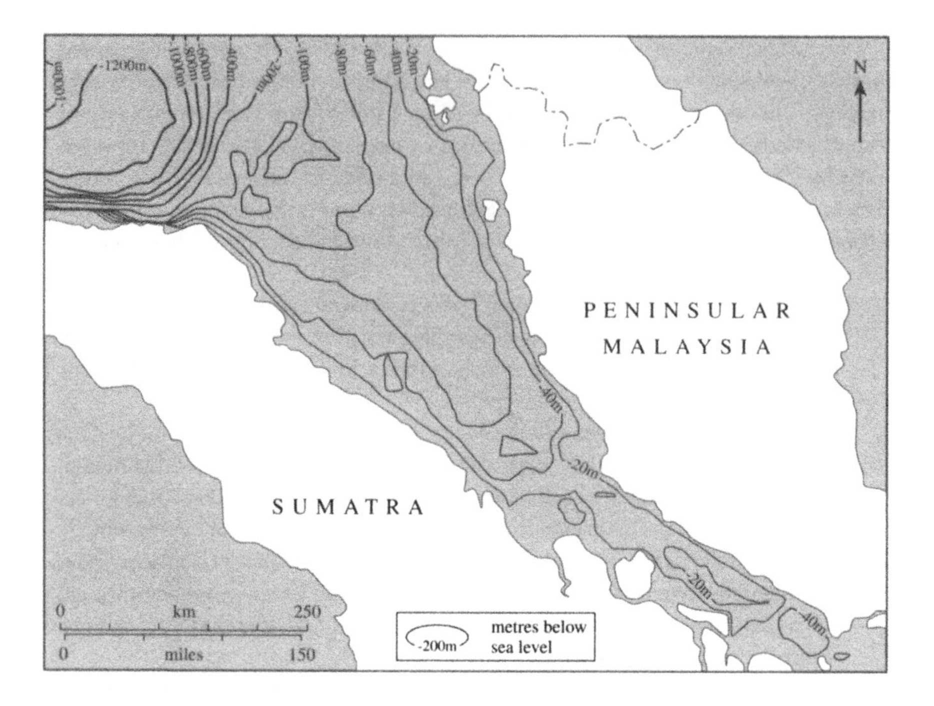

Figure 2.3 Submarine contours of the Straits of Malacca.

Figure 2.3, which shows the outline submarine contours of the Straits and Figure 2.4, which maps sediment types.

In Sumatra, the Barisan Range and associated mountains are the products of the oblique subduction of the Indo-Australian plate. So, too, is the continental-based volcanic arc which extends the entire length of Sumatra and which rises to an altitude of more than 4,000 m. A discontinuous line of northwest-trending straight river valleys and intermontane depressions is traceable for virtually the entire length of the Barisan Mountains, effectively defining the Sumatran Fault System, a series of dextral strike-slip faults which parallel the Sumatra trench, and lie some 300 km inland (Jobson *et al*, 1994).

The Barisan Range consists largely of upper Palaeozoic and Mesozoic sedimentary rocks that form the foundations of quaternary volcanoes. Acid volcanic products cover large parts of this pre-volcanic surface of the mountain. In the northeast of the geo-anticlinal upwarp a geo-synclinal hinterdeep occurs, where low hills and alluvial plains are located. The alluvial plains along the eastern coast become gradually narrower towards the north. Evidently sedimentation could not keep pace with subsidence in this area.

The Main Range and smaller ranges in Peninsular Malaysia have slightly different origins. Nevertheless, while the main processes producing the Barisan Range and the Main Range may differ, there are physiographic similarities. One example are the

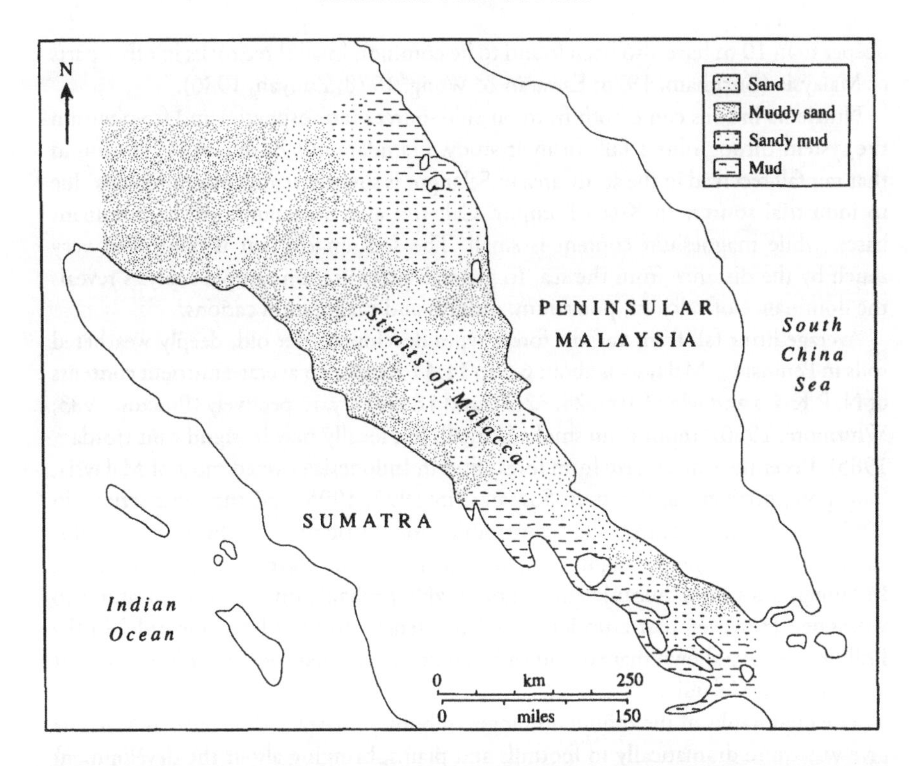

Figure 2.4 Sediment types in the Straits of Malacca (adapted from Keller & Richards, 1967).

granitoid provinces of Sumatra and Peninsular Malaysia. Isotopic composition of the granitoids west and east of the Semangko Fault shows that the granitoids of East Sumatra (east of the Fault zone), including the tin islands of Bangka and Belitung seem to be related to the older Central Granitoid Province of the Malay Peninsular (Gasparon & Varne, 1995), all rich in tin ore. The nature of the Straits' 'catchment' to a large extent governs the water flow and sediment transport into the main water body, and to the same extent determines the adjacent coastal plain evolution and development.

The lowlands and foothills below the western flank of the Main Range have been subject to weathering and denudation over long periods of time. Different rock types are found on this western part of Peninsular Malaysia right up to this mountain range. Much of the Main Range consists of granite intrusions of the Mesozoic, while schists of Middle Upper Silurian age have been found in some parts of Selangor. Given the wet and warm climatic conditions and the luxuriant rain forest vegetation, weathering processes have been effective for long time periods. A study by Hamdan and Burnham (1996) in parts of Selangor and Pahang showed that weathering depths of 27, 16 and 10 m, respectively, were found for granite, basalt and schist. Profiles deeper than 10 m have also been found to be common in all three rocks in other parts of Malaysia (Burnham, 1978; Eswaran & Wong, 1978; Zauyah, 1986).

Nutrients in soils come both from outside sources (atmospheric) and from within the system through litter fall. In their study, Hamdan and Burnham (1996) found that rainfall received in the study area in Selangor is slightly acidic, most probably due to industrial sources in Kuala Lumpur. Calcium and potassium are the dominant bases, while magnesium content is small. The sodium content is controlled very much by the distance from the sea. In general, precipitation sample analysis reveals the dominance of calcium, potassium, sodium and magnesium cations.

Average litter fall from various forest types developed over old, deeply weathered soils in Peninsular Malaysia is about 6.5 to 11.4 t/ha/y, with average nutrient contents of N, P, K, Ca and Mg of 100, 28, 32, 70 and 18 kg/ha/y, respectively (Proctor, 1983; Whitmore, 1990). Input from smoke is small, but locally may be significant (Jordan, 1985). Recently, smoke haze from forest fires in Indonesia covered most of Malaysia, Singapore, northern and eastern Sumatra (in 1992, 1995, and the most serious in 1997) for several months. The contribution of nutrients into the soil from this source must certainly be significant, though no studies have been conducted yet. In Sumatra, input of volcanic ash is more highly probable due to closer locations to volcanoes. However, past eruptions of great intensity from as far as Pinatubo in the Philippines have shown that volcanic ash can be transported over long distances as far as Peninsular Malaysia.

The high flanks of the mountain ranges in both Sumatra and Peninsular Malaysia give way quite dramatically to foothills and plains, bringing about the development of fast flowing river systems in an almost east–west direction into the Strait. These rivers used to act as efficient sediment conveyor belts to the coast, but are now undergoing much modification such that their role as efficient transporters of sediment has been weakened.

COASTAL PLAIN DEVELOPMENT

Coastal plains are a characteristic feature of the physical and human landscape. In eastern Sumatra such plains may extend between 48 to 193 km wide, while in western Peninsular Malaysia they are narrower with an average width of 64 km. These plains are geologically young, having started to form only in the last few thousand years (Woodroffe, 1993) and they contain primarily Quaternary and Recent sediments. As we noted earlier, for most of the Quaternary, Sundaland, which includes the Straits of Malacca, was one land mass, subject to sustained subaerial exposure. By the Holocene, the worldwide rise in sea-level had inundated much of the area and it was during this time that the coastal plains in eastern Sumatra and western Peninsular Malaysia evolved.

The processes of coastal plain development in Peninsular Malaysia can be inferred from a study of the Kuala Kurau mangrove shoreline of Perak (Kamaludin, 1993). Based on bore hole samples and palynological analyses, Kamaludin concluded that the coast has expanded significantly, the widest stretch being some 28 km. This process started some 5,000 years ago when the sea-level was at its highest, about 5 m above the present level. Coastal plain development occurred as the sea receded to its present level. The process of accretion through sediments supplied from the major river systems did not continue at a constant rate. Recent records suggest a trend of coastal plain development. From 1914 to 1969, map and air-photo evidence points to accretion rates of 1.8 to 5.4 m per year along certain sections of the coast, accounting for some 26.7 sq km of mangrove forest during this period. Understandably, this increase in accretion could be human-induced as agricultural activities like planting of sugar cane and paddy and, in particular, tin mining during this period, had caused widespread erosion, transportation and sedimentation along the coasts of Perak. The period from 1969 to 1986 saw a loss of 0.5 sq km of the coast. Coastal erosion has caused not only this part of the Perak coast to be eroded, but others along the Malacca Straits as well. The National Coastal Erosion Study (1985) conducted from Sungai Kerian to Sungai Gula, a distance of 39.3 km, reported a 33.4 km length of retreating coastline.

The reasons put forward to explain this contemporary coastal retreat include reduced sediment supply by rivers along the west coast of Peninsular Malaysia within the last two decades. This reduction can be attributed to several causes such as the cessation of mining in much of the hinterland, building of dams in the upper reaches of large rivers for flood prevention and hydro-electricity generation, and human exploitation of mangrove swamps too close to the shoreline.

Peat deposits dominate much of the eastern coastal plains of Sumatra. Peat lands, which are typical of many of the coastal plains of Southeast Asia, share many similar characteristics and processes of development. Taking Anderson's (1964) model, coastal mud and sand flats are initially stabilised by mangrove vegetation and later inhabited by freshwater swamp forest. The latter ecosystem is subject to flooding and sediment influx. Flooding provides nutrients and oxygen for luxuriant plant growth, but also promotes surface decay from insects, fungi and microbes. The accumulation of initial humus provides a high ash and, in brackish-water influenced areas, high-sulfur peat, upon which successive vegetation grows.

In the alluvial plains of Sumatra several fault lines can be distinguished. In South Sumatra they have a WNW–ESE direction, whereas in the alluvial plain of Central Sumatra the direction is parallel to the Barito Range. In this area, the peneplain between the Barito Range and the alluvial plain shows a few large and a number of smaller fold axes (anticlinoria and synclinoria). These more or less weak structures (fold axes and fault lines) often control river courses.

Towards the north the alluvial plain becomes narrower, and hence the slopes of the rivers increase, resulting in higher rates of transport of sand and silt to the sea. Northwest from the Sungai Wampu mouth, about 50 km northwest of Medan, the alluvial plain is very narrow and consists mainly of beach ridges. In their study of the peat area of the deltaic plain of the Batang Hari River in Jambi, Sumatra, Esterle and Ferm (1994) show that the presence of clayey sapric and fine hemaic peat throughout the upper half of the core suggests that this deposit was continually subject to flooding for at least the last 2,300 years of peat accumulation. The coastal plains of Central and North Sumatra share similar characteristics to those in Jambi.

CLIMATE AND HYDROGRAPHY

The climate affecting the Straits of Malacca is governed by the equatorial monsoon conditions that affect the region. By virtue of their location relative to the equator these areas are subject to the weather patterns prevailing within such a belt, marked generally by permanent low-pressure systems and the seasonal movement of the Inter Tropical Convergence Zone above and below the equator. However, the locations of Peninsular Malaysia and Sumatra and their orientations relative to the two monsoon wind systems determine to a large extent seasonal and annual distribution of rainfall in these two regions (Figure 2.5).

Rainfall distribution is further influenced by the topographic features of the Main Range in the case of Peninsular Malaysia and the Barisan Range in the case of Sumatra. Their almost parallel location and the northwest–southeast alignment blocks the movement of the monsoons, the former blocking the northeast monsoon winds and the latter the southwest monsoon winds. Winds are not normally pronounced in this region. During the northeast monsoon, from December to March, when the Inter Tropical Convergence Zone is farthest south, air from the north tends to be steady and moist. During the southwest monsoon, winds are typically light. For most of the transitional period, winds are light and variable.

The windward slopes receive more rainfall during the season when the winds blow on-shore. During the southwest monsoon, the eastern parts of Sumatra are on the lee of the Barisan Range and thus receive less rainfall than the western parts. Similarly, during the northeast monsoon, the western parts of Peninsular Malaysia receive much less rainfall than the eastern parts because of the influence of the Main Range. This orographic effect of the ranges extends beyond the immediate locality to the Straits themselves. However, spatial rainfall distribution in the Straits is difficult to construct because of the absence of any network of rainfall stations.

While it can be surmised that the presence of orographic features on lands bordering the Straits has some influence on the rainfall received, it is more difficult to gauge the influence of the Straits on the climate of the adjacent territories. Certainly, because of their narrow width, their influence as a major source of moisture for adjacent lands is not significant. However, on a localised scale, they have an influence on the development of line squalls or 'sumatras', which bring short but intense showers in the early mornings during the southwest monsoon months along the southern half of the western coasts of Peninsular Malaysia (Watts, 1954). Sometimes severe squalls are accompanied by gale force winds but, due to limited fetch, they have little influence on high wave formation in the Straits. The Straits are also significant in influencing land and sea breezes. In fact, there seems to be a clear relationship with land breezes in the west coast of Peninsular Malaysia initiating such squalls.

The influence of topography is more discernible during the limited monsoon

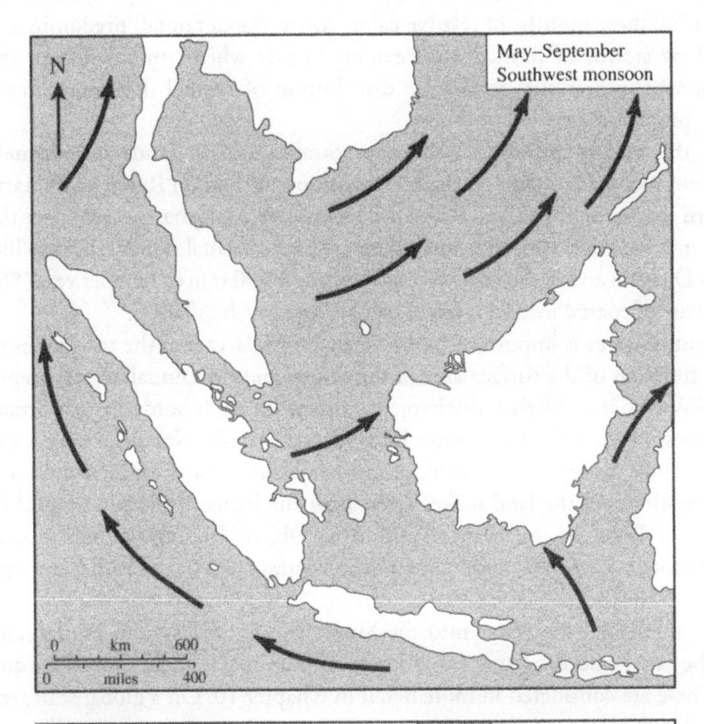

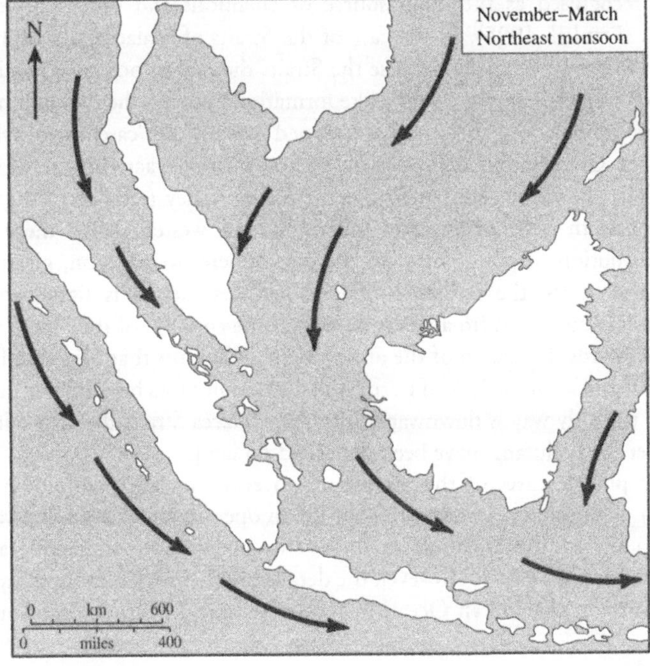

Figure 2.5 Monsoon wind regimes.

season. However, during the rest of the year, orographic factors do not exert any significant influence on the rainfall distribution in Peninsular Malaysia or in Sumatra. During these months of relative calm, convectional rainfall predominates, characterised by storms of limited areal extent. In fact when combined with the monsoon months into annual totals, the distribution of rainfall does not show a strong orographic influence.

Although the rainfall received within the Straits and on lands immediately bordering them may be less than on the western slopes of Barisan Range of Sumatra or the eastern slopes of the Main Range of Peninsular Malaysia, nevertheless the annual totals received in the intervening region are high. Coastal rainfall stations like Penang, Port Dickson and Malacca receive as much as 2,000 mm of rain per year. The same is also true of coastal locations in eastern Sumatra such as Jambi.

Direct input of water is important to the Straits of Malacca and the total amount received is a function of the surface area of the Straits and the annual average depth of rainfall. However, it is the river discharge coming from lands bordering the Straits that contributes much of the fresh water into this water body. Not surprisingly, the mountains and hills in Sumatra and Peninsular Malaysia increase the rainfall amounts greatly and the slopes of the land makes rivers flow efficiently to the sea. Places like Maxwell Hill in Perak or the Kedah Peak in Kedah, or Brastagi in Sumatra, at elevations of more than 1,000 m are some of the wettest parts of the Peninsular and Sumatra.

The direct input of fresh water into the Straits comes largely from rainfall and rivers, but the latter also discharge other kinds of inputs, particularly organics and chemicals. These are considered in more detail in Chapter 10. On a global scale, the atmosphere is recognised as the main source of chemicals and other pollutants into the oceans (Frankel, 1995). In the case of the Straits of Malacca, atmospheric pollutant input is not great chiefly because the Straits themselves occupy a relatively small area. Nevertheless, being in a valley-like formation between the two mountain ranges, the atmosphere above the straits is 'trapped' within this 'catchment' which may be significant. Given the greater industrial and other human activities in the west coast of Peninsular Malaysia, particularly in the Klang Valley (relative to the east coast) and the eastern coast of Sumatra (relative to the western part), there is a significant contribution of pollutants into the atmosphere. In addition, given the catchment of the straits, the influence of land and sea breezes is significant in maintaining prolonged to and fro air motion within the confines of the Straits with little tendency towards dispersion of the atmospheric pollutants that originate from both adjacent lands. Given the lack of horizontal dispersion of such pollutants out of the catchment, the only way is downwards into the Malacca Straits through rainfall or via rivers after the pollutants have been deposited on land.

The hydrographic features of the Straits are governed by high ground in the constricted part of the channel in the south, with an opening up to a sloping valley and finally delta towards the Andaman sea or the entrance from the Indian Ocean as discussed earlier (see Figure 2.2). However, the detailed bathymetry is more complex. As the Straits connect the Indian Ocean in the north and the Sunda Strait in the south, the influence of tides and currents as controlled by these two openings is discernible. The tidal characteristics in the Straits are semi-diurnal, with a tidal range between the lowest astronomical tide and the mean high water spring tide varying from 2.4 to 2.8 m (in Kuala Kurau). Whilst coastal processes are also affected by long shore currents and wave action, the latter is generally confined to more open sea. Low wave energy environments prevail in more sheltered estuaries and mangrove coastal areas.

The geological setting and earth processes experienced in the region have exposed the Straits to the effects of earth tremors and volcanic eruptions that originate in Sumatra. The Straits of Malacca, by virtue of their relatively shallow and narrow waters, do not experience tidal waves that may result from earth tremors such as to endanger vessels plying through their waters, nor to threaten coastal villages and towns. However, whilst experiences of earthquakes in Sumatra within recent times have not been severe, this does not mean that the Straits of Malacca are completely insulated from the effects of major earthquakes. Much attention has been focused on the environmental management of the Straits of Malacca these past three decades, but none of these explicitly include concern about earthquake effects on the Straits waters. Perhaps it is timely to consider the likely scenario of infrequent but probable major earthquake events and their consequences on life and activities within and in the periphery of the Straits of Malacca.

Certainly, past experience has shown that earth tremors have been felt in high rise buildings of major towns in Peninsular Malaysia and in Singapore, but no direct damage to buildings or loss of lives has resulted. Despite this, with greater awareness of the potential dangers and, particularly so, since the Kobe earthquake disaster in 1995, greater concern has been given to this matter by the Malaysian and Singapore governments. Continuing monitoring of seismic tremors and research is being undertaken to study such potential impacts as well as to re-examine the design and construction of high-rise buildings.

Current global concern about sea-level rise has some bearing on the Straits of Malacca. The rise of sea-level of 2 to 3 m above the present level in the Straits during the Quaternary provided opportunities for coastal expansion through natural processes of sediment deposition and peat swamp colonisation and expansion. In more recent history, it has also influenced the rise of the Malay coastal states, the growth of ports and economies of the hinterlands up to about the fourteenth century, when these centres started to decline. It is believed that the decline in importance of many of these coastal ports and towns on both sides of the Straits about 600 years ago may have been due to the fall in sea-level, which caused the expansion of mangroves and the shallowing of coastal waters. Khoo (1996), relying on geomorphological, archaeological and historical records, concluded that the fall in sea-level led to the shallowing of the Kedah sea and to the demise of important ports like the Merbok and Kuala Kedah; increased sedimentation in the case of the Merbok estuary, which was once a large lagoon, was exacerbated by human agricultural and tin mining activities at the foothills of Gunung Jerai. The same might have been true of many important coastal states and inland centres in Sumatra.

The current concern, however, is not with sea-level fall but sea-level rise, which will bring in its train a whole series of implications. Greater economic and physical development near and on the coasts of the Straits mean that more areas are vulnerable to inundation and more investments are required to protect such vulnerable areas. This will be discussed in more detail in Chapter 10.

3

LANDSCAPES, VEGETATION AND SOILS

As Chapter 2 has shown, the Straits region shares a number of common physiographic characteristics which are reflected in the landscapes of the region. In terms of river catchments, the dominance of certain soil types and common vegetational patterns, there is justification in seeing the essential unity of the region. Whilst divided by the waters of the Straits, as the preceding chapter demonstrated, those waters arrived only very recently in geological time. The nature of the landscapes, drainage, soils and vegetation are examined here; they provide the important environmental framework within which human ingenuity, cooperation and organisation has acted to shape the human landscape of the region.

RIVER BASINS AND THEIR ROLE

The overall 'catchment' of the Straits of Malacca stretches from the Barisan Range of Sumatra on the west to the Main Range of Peninsular Malaysia on the east. It is within this basin-like feature, and flowing from the two divides, that the major river systems operate.

The topography of the lands on both sides of the straits comprises the steeper hill slopes inland, through to the gentler foothills and lowlands that finally merge with the coastal alluvial plains. On both flanks of the Straits two types of rivers can be distinguished, although the two types are more marked on the Sumatran than the Malaysian side. First, are the large rivers that originate in the upland or mountainous interior. Second, are those that drain purely the lowland areas (originating only in the low-lying hills or plains). Rivers in the first category are shown in Table 3.1. Generally having a catchment area of over 5,000 sq km, these rivers are characterised by gentle flooding regimes, high sediment loads, a fertile silt and, in the lowland plains, significant meandering. The latter is almost a defining feature of the Sumatran lowlands bordering the Straits.

The second group consists of rivers that have low gradients in their entire courses, much too low to be able to discharge peak flows resulting in extensive flooding along the rivers and their tributaries (eg. Air Lalang in Sumatra). Tributaries draining swamps and flat areas, especially those draining peat swamps, have no instantaneous

Table 3.1 Major rivers abutting the Straits of Malacca

Sumatra		*Peninsular Malaysia*	
Province	*Rivers*	*State*	*Rivers*
Aceh	Krueng Aceh Krueng Meureudu Krueng Peusangan Krueng Jambuaye Krueng Tamjang	Perlis Kedah Penang	Sungei Perlis Sungei Kedah Sungei Muda Sungei Perai
Northern Sumatra	Sungai Wampu Sungai Belawan Bahbolon Sungai Asahan Sungai Barumun	Perak Selangor	Sungai Kerian Sungei Perak Sungai Berbam Sungei Selangor Sungei Kelang Sungei Langat
Riau	Sungai Rokan Sungai Siak Batang Kampar Batang Indragiri	Northern Sembilan Melaka Johor	Sungei Linggi Sungei Melaka Sungei Muar Sungei Pahat Sungei Johor

peak flow. This is due to the very low gradient of the rivers, the spongy character of peat soils with enormous water holding capacity, and the presence of tree trunks and other vegetation that hampers drainage. Tributaries draining peat swamps generally have a low to very low pH (5 to 3) which makes the water unsuitable for domestic and (in most cases) irrigation purposes. The sediment load is very low.

If it is the large rivers that account for much of the annual water, waste and sediment discharges into the Straits, and are of most significance in influencing the biochemistry as well as the sediment content of the water body, smaller rivers are by no means insignificant in their influence on the coastal and marine environment. These small, short rivers generally flow from the narrow coastal belt that, in the case of the Malaysian coast, tends to be more highly developed and populated. Together they account for a significant input of biological, chemical and solid wastes into the straits.

In the southern part of the region, Singapore and the Riau islands constitute smaller sub-regions which are very much influenced by the small rivers that flow into them. Again, though smaller in size and discharge, they are no less significant in contributing fresh water flow as well as waste discharge into these narrow straits. In Singapore the largest river basin occupies only 26.3 sq miles, the rest are by any standard very small. Many of these urban rivers have been altered by diversions and artificial channelling to overcome problems of flooding which have historically affected some residential locations. Rivers in Singapore discharge much less water and waste into the Straits than was the case two or three decades ago, largely as a consequence of better management of the urban environment.

The major rivers and their basins have had a significant influence on the human ecology of the region. They have been the loci of settlements, particularly in the more accessible lowland plains and valleys. Flat land and good alluvial soils regularly replenished with silt deposited during floods, render such areas highly cultivable. Rivers also provide water for drinking, washing and bathing, and for crop requirements in addition to being natural sources of fish to supplement human protein needs. In the early development of the Straits region, when direct communication between river valleys was difficult due to rough terrain and thick vegetation, rivers formed the most convenient means of communication for settlements along and between river valleys, the latter via a longer route by exiting to the coast first.

In the choice and establishment of early settlements in the Malay Peninsula, river estuaries and strategic sites along river valleys were often the preferred locations of settlements. In some cases, these settlements, precisely because of their strategic locations, have grown to become foci to control the movements of people and goods up and down river and consequently they have grown into seats of local government. The imposition of taxes strengthened the control and influence of the ruling elite, many of whom subsequently became the chieftains and sultans. Not surprisingly, then, at such strategic locations some of these settlements consequently became royal towns and seats of government in pre-colonial Malaya. Towns like Kuala Perlis, Kuala Kedah, Kuala Kangsar, Kuala Selangor, Kuala Linggi and Johor Bahru in the Malay Peninsula owe much of their importance to such an origin, although most of them no longer hold such functions today. The corollary is, to a large extent, true of some parts of eastern Sumatra, although there the extensive width of the coastal plain made movement between river basins more difficult.

While the coastal areas were generally more attractive for settlements, the upland interior was left very much untouched in the historic past, occupied mainly by semi-nomadic aboriginal groups, most probably pushed inland by the stronger and more established coastal communities. With settlement came the need to cultivate food crops. Hence, it was in the western coastal belt of the Straits that paddy cultivation took place from early times. There is sufficient evidence to indicate that rice was introduced into the northwestern parts of the peninsula from Siam during the fifteenth century via the trading waterways (Taylor, 1981). In the Kedah plain, large scale rice cultivation began as far back as three centuries ago (Affifuddin, 1975) and by 1949 most of the present day Muda area was already under rice cultivation. Through a system of canals to drain excess water, and a network of dykes along the coast to prevent tidal salt-water intrusion, peat swamp areas were gradually converted to paddy areas. During the British colonial period large paddy schemes were initiated through better regulation of water by irrigation canals. Three important schemes were the Krian irrigation system in Perak, Bernam in Selangor and Muda in Kedah-Perlis. But the largest rice growing area was the Muda Irrigation Scheme, which encompasses much of Kedah and parts of Perlis and Province Wellesley, with a combined acreage of some 97,000 ha.

Subsequent economic development has tended to concentrate on the western coastal belt of the peninsula. Not only were large tracts of coastal lowlands converted for different uses including paddy cultivation and settlements, areas further inland were gradually developed. The construction of roads and rail rendered more inland areas accessible. Large tracts of forest land gave way to plantations and settlements. Following the introduction of cash crops such as rubber, oil palm and cocoa, large-scale forest clearance was embarked upon and large plantations for these respective crops, owned by foreign companies, were established. Smallholdings by local landowners also grew in number and total acreage. The pace of this change was more marked on the Malaysian than the Sumatran side, a reflection, in part at least, of the difficulties faced in reclaiming the extensive mangroves that flank large areas of the east coast of the Straits.

PATTERNS OF DISCHARGE AND DRAINAGE

Since the Straits of Malacca are a sink and receiving area of sediments, water and other discharges, changes in the river basin dynamics of adjacent lands have a direct influence on the ecology of this water body. One of the most significant influences on the ecology of the Straits in recent years has been the steady curtailment of discharges from the mainland. Changes in flow due to river regulation and dam construction have undoubtedly altered the salt diluting qualities of waters entering the Straits. On the Malaysian side, dams have been constructed at several points along major rivers for the purposes of irrigation, hydro-electricity generation and flood control as shown in Figure 3.1.

On the Muda River in Kedah for example, two dams were constructed in its upper tributaries (the Muda and Pedu) to provide irrigation water to the off-season second crop in the Muda Irrigation Scheme, the rice granary of Malaysia. The flow of the Perak River has been altered through the construction of the main Temmengor dam for flood control, with smaller structures on many of its tributaries. On the Klang River, the Klang Gates dam was constructed in the 1960s to provide drinking water to the Klang Valley. Since then, many more impoundments have been added on this river and on its sister tributary, the Gombak, as well as the Langat River, to meet the increasing demand for water by domestic and industrial sectors in the Kuala Lumpur–Klang conurbation. This trend will continue for as long as surface runoff is the main source of Malaysia's water supply.

Curtailment of discharge is not of major significance in the case of Sumatran rivers. However, a major dam, the Asahan, was completed in the late 1970s mainly for the generation of electricity, with a secondary purpose of flood control. Such a major project has reduced the pattern of annual river discharge significantly. The need for additional water supply to meet domestic and industrial demands means that more and more rivers are likely to be regulated in future. The impact of such regulation of water from rivers that historically flow freely into the Straits has not been studied. Whether the reduction in discharge has any significant negative effects

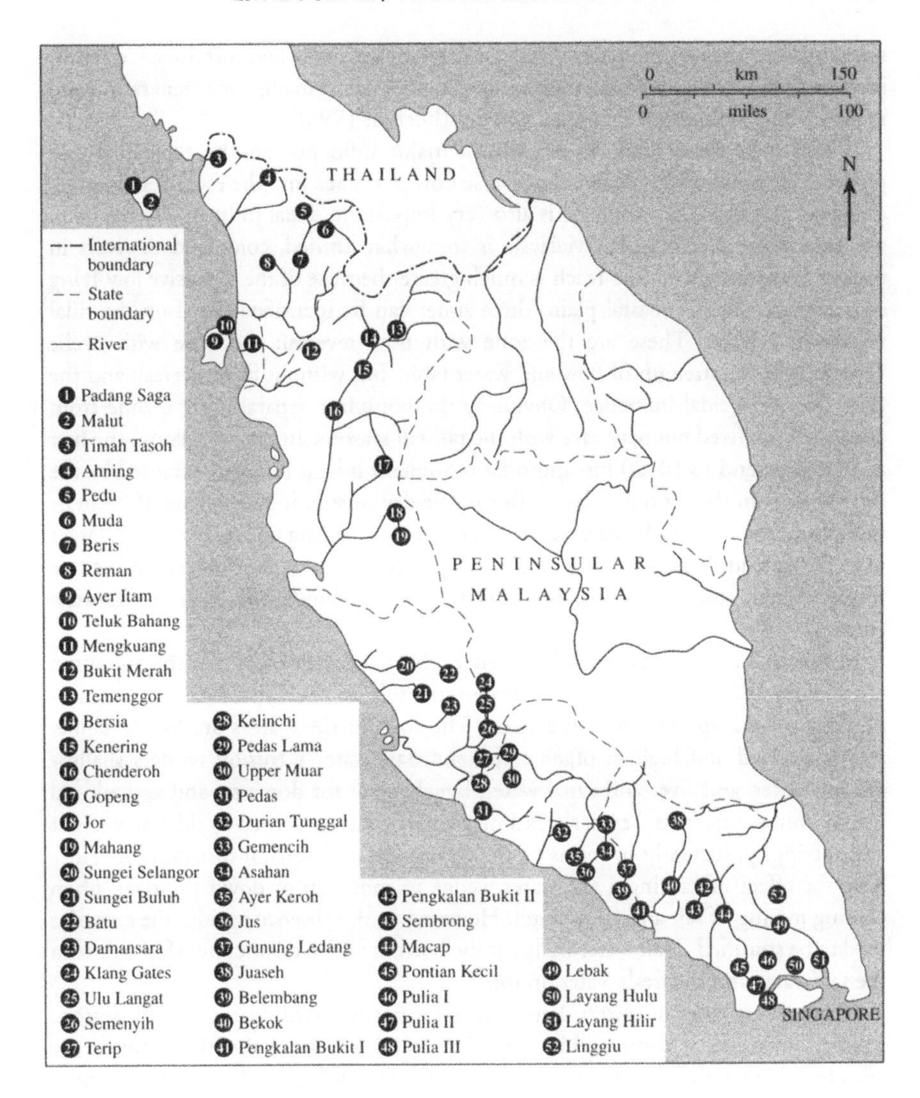

Figure 3.1 Dams in west coast Peninsular Malaysia.

on the marine ecosystem is conjectural. Since the Straits of Malacca are not an enclosed sea, a reduction of fresh water flow may be offset by this free flow along the Straits.

Perhaps of more significance than the curtailment of water discharge are the changes in the movements of organic and other materials from the land into the Straits which may affect marine ecosystems. The curtailment of fresh water discharge may result in the riverine and marine waters not meeting minimum environmental standards in terms of pH, heavy metal content, transparency and turbidity. Certainly, increased salinity beyond the estuarine areas would favour mangrove regeneration and increase the productivity of coastal fisheries (Nichol, 1993).

Whilst river flows into the sea have a major influence on the ecological and physical character of the Straits, the reverse flow of the sea into the estuaries and river channels through tidal influence is also very important. Tidal influence in rivers on the west coast of Peninsular Malaysia is somewhat limited, compared to rivers in eastern Sumatra where tidal reach is much greater because of the extensive low-lying coastal plain. In this coastal plain, three zones can be identified based on the tidal regime and flows. These are the zone with flow reversal; the zone with cyclic variations in the strength of flow and water table, but without flow reversal; and the zone free from tidal influence. Obviously the boundary separating one zone from another is not fixed but may vary with the rainfall seasons. In the wet season, the first zone may extend to 10–20 km and tidal movement inland is largely retarded by the river flow from the central uplands. But in the dry season, it may extend 100 km or more from the coast as the volume of tidal movement during one cycle is much larger than the upland discharge. Other factors that may influence the tidal extent include friction losses and river slope, but seasonal rainfall conditions exert the greatest control.

In this region of extensive tidal influence, characteristic of much of the Sumatran coast, tides and salinity play a very important role in the hydrology and geomorphology of swamp areas near the coast. The soils in these areas are highly saline, poorly drained and high in organic content. Salt water intrusion renders shallow ground water, and river and canal water, largely unfit for domestic and agricultural use. Within the rivers and estuaries, salinity is very much governed by tidal movement and mixing, particularly in terms of the horizontal and vertical distribution. Tides have the effect of causing a salt water wedge to move up or down the river, often causing mixing of salt and fresh water. However, in the absence of tides the estuaries tend to be stratified—salt water wedge at the bottom with a distinct interface between the salt water and the fresh water on top.

For most estuaries in eastern Sumatra, the water is mixed during the dry season. Salinity intrusion is considerable then. Up to 60 km from the river mouth the salinity is higher than 5,000 ppm, at 100 km it is 1,000 ppm; limits which render the water unsuitable for both domestic and irrigation purposes. In the wet season, however, some estuaries become stratified because of the much higher upland discharge and the relatively small tidal flood volume, while others remain mixed due to the very large flood volume. Salinity intrusion is not a major problem during the wet season because of the dilution effect of larger volumes of fresh water discharge.

VEGETATION

Four types of vegetation occupy the area between the foothills and the sea on both sides of the straits. These are the fresh water swamp forest, the brackish water forest, the coastal forest and the tidal mangrove forest. Of the four, the coastal swamp forest is the most significant so far as the Straits of Malacca are concerned. This forest type is facing immediate and serious threat of depletion.

The coastal swamp zone occupies a broad, flat and poorly drained area between the mountains or undulating plain and the sea. The coastal swamp zones may vary from just a couple of kilometres to more than 100 km wide and they result from a land-building process based on eroded upland materials being built out into the sea. Sand, silt and clay are colonised by vegetation, and successive floods deposit more organic and inorganic material. Since anaerobic conditions prevail, organic matter is rapidly accumulated and builds up the surface of the land. As the land extends seawards, the older swamplands become less saline and less frequently flooded.

Swamp forests have also colonised the wide estuaries, formed by the drowning of river mouths that have subsequently been partially filled with huge quantities of detritus deposited by the truncated rivers flowing into the shallow seas. The very process of drowning led to an increase of the rate of sedimentation and along many of the shallower and less disturbed shores of the archipelago, particularly on the Sumatran side, vast belts of swamps have built up.

The land-building process remains active as long as soil transportation from the upland continues. In eastern Sumatra, evidence of greatly accelerated erosion in the uplands of the interior can be seen. The colour of the streams betrays erosion as does the silting-up of rivers and the outward growth of the mangrove belt. Ports which in 1942 could be reached by small coastal steamers have lost their function as shipping points along the east coast of Sumatra (Pelzer, 1968). Many studies have shown sedimentation rates in mangrove areas ranging between 1 mm/a and 8 mm/a (Bird & Barson, 1977), implying their efficient function as land builders. A contrary view holds that mangroves are a result of sedimentation, not the cause of it, hence their role in coastal protection is questioned (Woodroffe, 1992).

The mangrove swamp ecosystem is one of the most productive ecosystems in the world and is an important spawning, nursery and habitat area for many economically important species of fin-fish and prawns. Its importance is reflected in the fact that 42% of fishery catches along the west coast of the peninsula, which has large areas of mangroves, are of mangrove-related species, while the corresponding figure for the east coast, which has fewer mangroves, is only 12% (Ismail, 1993). The importance of mangroves in performing other functions has been much discussed and these include trapping sediments and other particulate matter and helping to filter sediments and thus protect coral reefs and sea-grass beds. It also helps negate the damaging effects of waves and can help reduce coastal erosion. The stabilisation effects of the roots enable the ecosystem to advance towards the sea as more materials are deposited.

There are as many as 30 species of plants in mangroves including the *Avicennia, Sonnerratia* and *Rhizophora,* together with *Nipa frustican.* Mangrove forests in this region appear especially species-rich—in Sumatra as many as 25 different species are commonly found (Schwamborn & Saint-Paul, 1996). Mangroves can produce over 50 products such as food, paper, tannin, alcohol and dyes. Wood-chips, piling

Table 3.2 Mangrove forest decline in Southeast Asia

Country	*Period*	*Original area (ha)*	*Present area (ha)*
Peninsular Malaysia	1979–1986	113,000	89,000
Thailand	1961–1993	300,000	219,200
Vietnam	1969–1990	425,000	286,400
Indonesia	1969–1986	4,220,000	2,176,000
Philippines	1968–1995	448,000	140,000
Singapore	1922–1989	700	180

Source: Mastaller (1996).

poles, fuel wood and charcoal are produced from the swamp forest. However, much of this is unregulated and is characterised by small scale exploitation. The large scale clearing of swamp forests for development is discouraged because of their ecological importance as wildlife sanctuaries for both flora and fauna.

As Table 3.2 suggests, mangrove forests in the adjacent coasts of the Straits, like most mangrove systems in the region, are under constant threat of depletion for many reasons. The impact of wastes from land or sea can cause the ecosystem to undergo a slow dying process. In some localities they have been taken over by shrimp ponds, the development of new settlements, conversion to agricultural land, or just deterioration as a result of exploitation for wood and wood products. In the west coast of Peninsular Malaysia between 1965 and 1985 some 200 sq km of mangroves were alienated for other uses, from the total area of 1,184 sq km. This was despite the fact that the mid-1980s were economically depressed years for Malaysia. From the late 1980s the rate of destruction has accelerated. In Selangor alone, some 86 sq km were cut down between 1988 and 1989 (Collins *et al*, 1991).

In Singapore, mangrove forest is negligible as much of the coastal areas have given way to massive land reclamation for various purposes, such as the expansion of Changi airport, the naval base, houses and the construction of estuarine reservoirs to augment water supply sources (Wee, 1982). Patches of mangroves are found in the northeast coast of the main island and on a few off-shore islands such as Pulau Ubin and Pulau Tekong. Even in these areas, however, unless efforts are made to preserve the forests, they will be soon replaced by development. The Sungai Buloh mangrove area has been gazetted as a bird sanctuary, but in Pulau Ubin patches of ponds have appeared in some coastal localities for fish and shrimp culture. Attempts have been made recently to regenerate mangroves on the island.

The various purposes to which such conversions of mangrove forests were put have in turn brought about their own sets of problems. Conversion to agriculture can lead to soil acidity and crop failure, and the financial resources invested do not always justify the cash returns from crops produced. Toxic washouts often affect other adjacent ecosystems, resulting in the reduction of fish and shrimp populations and, in some cases, diseases. Ponds constructed in mangrove areas can result in strong acidification of pond waters, not to mention the limited options left for future use of the same area. In Sumatra, large scale and often illegal logging activities have caused major damage to the ecosystem. Much of the wood is used for paper and chipboards, and the setting up of such paper mills and chipboard factories produce discharges that are themselves harmful to the ecosystem. Damage is also caused by oil wells and oil extraction in the Dumai area of Sumatra.

SOIL RESOURCES

On the steeper slopes of the Main Range and the Barisan Range, soils derived from extensive granite weathering are formed. These reddish-coloured materials may be more than 10 m thick and may contain core boulders. They are generally acidic, clayey and low in nutrients, but the nutrient status depends to a large extent on whether the parent granite materials are fine grained or coarse grained, the former having the higher nutrient status. The more accessible areas where such soils are found tend to be cultivated with rubber.

In the highlands of Sumatra, fertile soils, the andosols, derived from basic volcanic rocks do occur in several places. Such areas are able to support settlement and intensive agricultural production and, because of their higher potash and phosphate content, and their deep and friable nature, they are suitable for many types of crops. However, there are many more areas in the mountains where the soils are infertile and do not support a high population density. For Sumatra as a whole, infertility of soils is a problem and leaching is high. In many areas traditional rice growing is often supplemented by the slash-and-burn economy on hill slopes, and more recently hill rice has been replaced with tree crops and fruits. Spices such as cinnamon are grown on the slopes of Mount Kerinci and Mount Merapi at heights of up to 1,000 m, while pepper can be grown on the eastern slopes of the Barisan Mountain in the south of the island. Temperate vegetables are also grown in the highlands to meet demands from urban and petroleum centres. Some tectonic basins and trenches within the highland zone filled over time with alluvial material support agriculture and settlements. A fault-block depression stretching through the mountains and lying in a northwest and southeast direction are home to the Gajo people in southern Aceh and to the Orang Lampung in the Ranau depression (Donner, 1987).

Soils derived from sedimentary rocks are found in Peninsular Malaysia, Sumatra and Singapore. In the west coast of the peninsula this type of soil is extremely variable in composition and nutrient status. While most of the soils are generally low in fertility, their structure renders them most suitable for agricultural development. This is true for west Peninsular Malaysia and eastern Sumatra, but in Singapore such areas are no longer used for agriculture as they have been taken over by residential and urban development.

Soils derived from limestone are present from Perlis to Selangor but they are limited in area. This is because while limestone geology covers much of these areas, the calcareous materials are generally covered by an overburden of non-calcareous materials brought down by rivers. Areas in Sumatra with this type of soil are also limited for the same reason. The high rate of leaching has been conducive to the formation of laterites, particularly in the better drained lowlands and foothills, mainly due to the exposure of oxides of iron and magnesium to the free oxygen in the air. Apart from the natural processes of leaching and subsequent drying, forest clearing has further enhanced this process. In general, because of the poor nutrient status and poor structure of the materials, laterites are not well suited to agriculture and cash crops are only possible with the use of artificial fertilisers.

It is difficult to provide an accurate classification of swamp soils in eastern Sumatra due to the complex geomorphology of the region. A rough classification based on composition, whether mineral or organic, is possible. Lobbrecht *et al* (1985) classified mangrove soils into three types: marine soils consisting of beach ridges, potential acid sulphate soils and acid sulphate soils; riverine soils comprising levee and backswamp soils; and finally organic soils which include peat, muck and organic clay. Others have classified the swamp soils into three major groups—entisols, inceptisols and histosols—each varying in chemical composition and fertility. Entisols are generally fertile, but high in sodium content and unripe organic matter, and therefore unsuitable for agriculture. Inceptisols contain more mineral soils and are fertile with less sodium content but more ripe organic matter than the former. This type of soil has potential for agriculture. Histosols or organic soils are mostly found between rivers in saturated and indurated conditions, with the lowest sodium content. The potential for agriculture depends primarily on the state of development and depth of the organic material.

Beach ridges comprising mainly coarse sand, loamy sand and sandy loam are relatively rare in the plains of eastern Sumatra, because the gradient of the rivers is so small that the sandy proportion of the river load never reaches the sea but is deposited upstream. In Central Sumatra and the northern part of South Sumatra beach ridges of any importance are also lacking. An exception is in the narrow alluvial strip of the extreme south of Sumatra where beach ridges of 5–10 m above mean sea-level are found. They represent the effect of either the recent lowering of the sea-level or tectonic movement. Other beach ridges on Sumatra are found on the islands along the central part of the east coast (Tebingtinggi, Bengkalis, Rupat) or in the alluvial plains of the east side of North Sumatra. Beach ridges are more common on the west coast of Peninsular Malaysia where place names with the word prefix *permatang* are frequently used. Such beach ridges have been described in Chapter 2 as evidence of sea-level change in the recent past.

Marine clay soils originate from sediments deposited on saline, tidal flood plains and in the extensive mangrove swamps along river branches and tidal creeks of the alluvial plains of East Sumatra. In these areas conditions for the formation of pyrite ($FeS2$) are very favourable, and in combination with acid volcanic rocks are conducive to the formation of acid sulphate soils. After oxidation of pyrite, which produces sulphuric acid, the pH of these soils becomes very low and this can result in toxicity which can damage rice and other crops. After drainage (natural or artificial) of potential acid sulphate soils, the physically unripe soil will crack and air will penetrate. In this well aerated but moist soil, pyrite is oxidised to dissolved ferrous sulphate and sulphuric acid. Bacteria increase the rate of oxidation and they also oxidise Fe^{2+} to Fe^{3+}, which in turn rapidly oxidises pyrite.

In more inland areas, river levee soils are found where riverine sediments are deposited on the potential acid sulphate soils of marine origin. This type of soil can vary widely from silty to loamy to sandy in more upstream levees of slightly higher topography, and is generally well drained. In the downstream direction, the levees become less pronounced and the soils become finer. In the homogenous swamp area, the levee soils are fairly firm grey or grey brown clays. The soil profiles tend to be more mature. The backswamp soils tend to be very fine or fine (heavy) clays. The clays are often sticky when wet and colours vary from grey to grey brown depending mostly on organic matter content. All backswamp soils are poorly drained. The fertility of these soils depends very much on the natural fertility of the riverine silt.

The organic soils classified as peat, muck or just organic are differentiated on the basis of loss on ignition—the greatest loss of 65% is for peat and the least (25–35%) is for organic; while the loss on ignition is somewhere in between for muck. All three types are found in eastern Sumatra. As soon as the frequency of tidal flooding of a part of the alluvial plain is reduced to a few times per year, the brackish water vegetation of *nipah* palms changes into fresh water swamp forest. This fresh water swamp forest delivers the organic material for the formation of organic soils.

The type of organic soil most likely to develop depends on the distance to the nearest river and on the characteristics of that river. Along most rivers extensive flooding occurs during the wet season. This brings about the formation of natural levees and floods transport silt into the backswamps, where organic soils are developing. Thus organic material mixes with mineral soils. Depending on the period and frequency of flooding, organic clay, followed by river muck and, eventually, thin layers of peat are formed. The result is a soil with a mineral marine sub-soil under a layer of organic clay, which is covered by muck and/or thin layers of peat. These swamps located in the neighbourhood of rivers are called fresh water swamps and the soils called eutrophic, shallow peat or Fluventic Topofibrist or Terric Tropohemist. Further away from the rivers, where no flooding occurs, true peat swamps develop. In these areas, drainage is impeded by tree trunks and other organic materials, resulting in a permanent waterlogged condition. Under these circumstances thick layers of peat, or so-called peat domes develop. These peat soils are called oligotrophic, deep or dome peat and they have a very high content of organic matter (up to 97–99%).

The thickness of the peat layer can be 10 m or more (13 m have been reported on Sumatra), but near the coast the layers are thinner. The materials underneath the dome are spongy and able to absorb large quantities of water so that the peat contains at least 90% water in volume, the solid mass being formed by tree trunks, branches and roots in various stages of decomposition. The remaining finer smeary particles are composed of other plant debris from tree bark, mosses, ferns and other vegetation derived from the under stories of the peat forest. All peat soils have a low pH: below 5 for the shallow fresh water swamp peat and generally below 4 for the deep, oligotrophic peat.

PEATLAND EXPLOITATION

The importance of peat soils depends on their suitability for agriculture (normally for rice or soya in case of dry season crops), which in itself is not easy to classify. Soil suitability classification depends on soil and water conditions which influence the fertility or toxicity (acidity and salinity) or structure (ripening and bearing capacities, possibility for tillage, tree trunks) of the soil. In terms of water conditions the suitability of the soils will depend largely on the possibilities for drainage and irrigation. Deep oligotrophic peat soils, coarse sandy soils and mangrove soils are generally considered unsuitable for rice cultivation. Sandy soils are low in fertility and have poor water retention capacity, while mangrove soils (halic sulfaquents) are regularly flooded with saline water and contain sulphuric material within 0.5 m of the surface.

Peat soils are much more extensive in eastern Sumatra than in the west coast of the Malay Peninsular. The development of such soils for agriculture has not been widely advocated because of the inherent problems of low productivity; most observers consider them as being of no more than marginal importance (Brown, 1972). But pilot projects in Indonesia have indicated that as soon as the peat land is connected to a river and the leaching process is allowed to take place for around three months, the land is ready for rice cultivation, and the first rice crop can be expected after one year (PUTL, 1974). In reality, the success of agriculture on peat soils in Indonesia has been highly variable, and crop yields on average have been low (Radjagukguk, 1992). Despite such results, the need to open up more land to sustain Indonesia's policy of self-sufficiency in rice has increased pressure to use peat land resources for agriculture. It is evident that with high population pressure and a scarcity of arable land, marginal soils like peat will have to be opened up. While much of Indonesia's peat land is found in Kalimantan (50.4%), peat land in Sumatra accounts for some 25% of Indonesia's total peat acreage (see Table 3.3). Swampland development is increasingly important in the environmental transformation of the region (Hanson & Koesoebiono, 1979).

Peat soils in Jambi and Riau are generally deep, whereas in South Sumatra they tend to be shallow to medium depth. In general, peat soils of less than 2 m are considered suitable for wetland rice, while deep peat is not. It is characteristic of peat soils to have low pH (ombrogenous peat has values of 3–4), and frequent burning of surface vegetation can increase the pH, an activity that is common in these areas during dry spells. If the depth of peat is suitable for rice, with proper control and management of water, such soils can be highly productive.

Peat lands in Sumatra as in other areas of Indonesia have been opened up by transmigrants for horticultural crops such as vegetables and fruits, particularly those near population centres, irrespective of peat thickness. Intensive management, controlled burning and the application of fertilisers such as decomposed fish residue, ashes of rubber residue and organic debris and liquid animal manure can lead to high profitability. The need for animal manure makes it essential for farmers to integrate animal husbandry with cultivation of crops.

Plantation activities in the interior have also increasingly shifted to lowland peat

Table 3.3 Peat swamp forest in Indonesia

Province	*Extent (× 1000 ha)*	*% of total Indonesia*
Aceh	270	1.5
Northern Sumatra	335	1.8
Riau	1,704	9.2
Jambi	900	4.9
Southern Sumatra	990	5.4
Total	4,199	22.8

Source: Chua *et al* (1997).

areas of eastern Sumatra, particularly in the Riau Province where cash crops like coconut and oil palm are grown, with possible prospects for coffee, cocoa and rubber growing on this type of soil. Such large scale plantation necessitates the construction of effective drainage systems and the utilisation of deep peat where upon compaction the use of heavy machinery is possible. Industrial crops like ramin (*Boehmeria* sp.) and some spice and medicinal crops also grow quite well on peat here.

Large parts of the eastern plains of Sumatra remain as yet unreclaimed, very sparsely inhabited and are still covered by mangrove swamp. These areas were traditionally seen as having the potential to solve the problem of landlessness in the overpopulated islands of Java, Madura, Bali and Lombok. The problem became more urgent after the Second World War when the populations on those islands continued to grow. The transmigration programme which was expanded from the late 1960s sought to encourage large numbers of people to migrate and open up lands in these low-density areas (Cleary & Eaton, 1996).

Reclamation of land took two forms—one encouraged by the government and the other, the spontaneous migration and reclamation of mangrove lands. In Sumatra, pockets of peat swamp were converted to agriculture through a slow process of cutting the forest, controlling water supply, improving the soil condition and stabilising the land. Interviews with settlers in Jambi reveal that a number came on their own some 25 years ago, and the land is not completely reclaimed yet. Farmers had to dig small drainage canals, clear the land and construct small dykes to keep water on the fields. Such areas have been turned to paddy fields (*sawah*), while on slightly higher ground or where wider main bunds have been constructed, coconut and some fruit trees are planted. Despite the fact that spontaneous migrants had to rely completely on their own initiative and resources, in terms of the overall land reclaimed, their efforts are as important as the government sponsored projects. Nevertheless, the overall achievement of the government's transmigration programme over the years has been paltry. The impact of land clearing in Sumatra would have been more severe had the programme achieved a greater level of success.

In Peninsular Malaysia peat varies in depth; 9 m has been reported (Mohd Zahari *et al*, 1981), but overall peat soils are thinner than in Sumatra. Much of the peat soil in Malaysia is found on the coastal lowlands lying between the lower stretches of the

Table 3.4 Peat swamp forest in Peninsular Malaysia

State	*Extent (× 1000 ha)*
Johor	228
Negri Sembilan	6
Selangor	194
Perak	107
Pahang	219
Trengganu	81
Kelantan	7

Source: Chua *et al* (1997).

main river courses, flanked by low-lying incipient levees (Table 3.4). On the seaward side, they grade into mudflats or sand ridges. Towards the interior, they transgress into low hilly country blanketing some of the small rises and covering some of the valleys (Tie & Lim, 1992). These are generally poorly drained depressions or low-lying coastal plains.

In the coastal areas bordering the Straits, large areas of swamp and peat land have been developed to a more advanced stage than in Sumatra (Rieley, 1992). The more important peat areas are found in west Johor, northwest Selangor and in the trans-Perak areas in the central and lower Perak districts. The soil consists of marine clay deposits, which may be sulfidic in nature or may contain an admixture of marine and riverine deposits especially along river courses. Generally the west coast peat has a high organic content with an ash content of less than 10% and a loss of ignition value exceeding 90% (Mutalib *et al*, 1992).

While peat areas can be made cultivable after being drained, one effect of draining of water and the decomposition of its intrinsically high organic content is the significant subsidence of the ground surface. Even after water is drained, through large investments in the construction of canals, ongoing problems of subsidence may render some of these areas water-logged. In the first two years of draining the peat in the West Johor Integrated Agriculture Development Project, a rate of subsidence of 50 cm was noted (Welch & Mohd Adnan, 1989). Even when the problems of drainage have been tackled, cultivation of crops has proved to be problematic. It appears that the three-layered sapric–hemic–fibric morphology is essential for successful cultivation of crops (Mutalib *et al*, 1992). Root growth and anchorage is encouraged by the sapric and hemic layers which is absent in the fibric materials. A necessary measure to obtain this morphology is for the peat to be drained, to encourage decomposition of the peat, resulting in the consolidation and compaction of the broken down organic materials, leading to the formation of the sapric and hemic layers. Other problems include the highly acidic nature of the soil, its low bearing capacity that prohibits the use of heavy machinery, and the presence of logs and tree trunks that prevents the mechanical cultivation of peat areas. Large-scale development of peat soils would be economical if mechanisation was possible, but the constraints to mechanisation on peat may render large projects uneconomical (Ooi,

1992). As far as the growth of crops is concerned, given proper treatment peat soils are highly suitable for certain crops such as oil palm, pineapples, vegetables, tapioca, banana and others; crops that are characterised by shallow rooting and fibrous root systems.

By the mid-1980s, some 313,600 ha of peat in Peninsular Malaysia had been developed for agriculture, accounting for some 32% of the total peat area. The majority of this area was in the west coast, abutting the Straits (Table 3.4). The major crops grown include oil palm, rubber, coconut, paddy, pineapple and mixed horticulture. The major proportion of 133,000 ha of oil palm is found in Perak, Selangor and Johor. The other major crops of rubber, coconut and pineapple are predominantly grown in Johor, with areas of 73,300 ha, 28,200 ha and 14,000 ha, respectively. Most paddy areas are found in Perak and Selangor in the west coast.

Probably the best example of large-scale conversion of peat land into agricultural use in the west coast of Peninsular Malaysia is in the West Johor Integrated Agriculture Development Project. Developed in two phases, it was launched in 1974 and completed in 1988; phase one occupying an area of 148,500 ha and the second phase, 228,000 ha. Within this project some 25% of the total acreage comprises peat land. The whole project has attempted a comprehensive agricultural development programme with targets set for complete implementation for the two phases: 2003 and 2016 for Phase I and Phase II, respectively. Among the crops grown are oil palm, rubber, coconut, pineapple intercropped with coffee, banana, tapioca and vegetables. But for the peat area, oil palm and pineapple predominate. Being a large-scale project with government assistance, much effort and money has been invested to develop the basic drainage and communication infrastructure. In the peat areas the provision of a proper drainage system and a relatively close network of unpaved agricultural roads has proved costly (Lim, 1992).

As pressure on land becomes greater in Peninsular Malaysia and to some extent the Riau islands, the coastal plains and peat areas are giving way to agriculture, while some agricultural area and swamp lands have been and will be converted to urban, residential and port development. Unless concerted efforts are made to conserve designated swamp areas, these ecologically important ecosystems will disappear altogether. In eastern Sumatra, however, the large expanse of peat soils make them attractive for new settlers to develop. However, the rate of depletion so far has been slow precisely because of the intrinsic difficulty the natural environment poses to land development by new settlers (see Hanson & Koesoebiono, 1979).

In the light of the increasing reclamation of swamp land in Sumatra and the deterioration of such ecosystems due to human exploitation, urgent attention has been drawn to the importance of this ecosystem in the overall economic and environmental character of the region. There is also debate as to which brings greater benefits to human populations: reclaiming the mangrove swamps or leaving the ecosystem in its natural state. In the past, activities that were seen as bringing direct benefits include the expansion of agricultural land especially for paddy cultivation, aqua-culture, and exploitation of the forest products by the informal sector. However, benefits derived by protecting swamp forest habitats and allowing them to carry out their ecological functions such as flood control, tidal storm prevention, groundwater recharge, coastal erosion prevention and fish breeding ground provision may prove to be greater. When properly measured, the total economic value of a wetland's ecological functions, the services and its resources may exceed the economic gains of converting the area to an alternative use (Barbier, 1993).

As this and the preceding chapter have shown, in terms of geology, morphology, soils and vegetation, the Straits region shares many common features. The distinctive nature of the drainage system, coupled with characteristic soil (peat) and vegetation (mangrove) types has created common potentials and, with the impact of economic and social development, common sets of problems. The ways in which those resources—organic, marine, mineral and human—have been exploited in both the past and present are the subject of Chapters 4 and 5.

Part 2

RESOURCES AND TECHNIQUES

4

RESOURCE USE IN THE COASTAL AND MARINE ZONE

The geological and geomorphological history of the Straits region has provided conditions under which a range of vegetation, landform and soil types have developed. The presence of important mineral resources—notably tin and hydrocarbons—also reflects geological conditions in the region. This range of resources has been instrumental in shaping the imprint of human settlement in the region. The nature and distribution of both land and marine resources has, together with the impact of technological change, shaped the pattern of development in the Straits. Thus, the presence of hydrocarbons in the sedimentary basins of Aceh and Dumai in eastern Sumatra, important tin ores in the river valleys and coastal plains of the western parts of Peninsular Malaysia and the islands of Banka, Singkep and Billiton in Indonesia, and kaolin and ball clay in the tin areas of Malaysia, especially in Perak, have all left their imprint on the historic and contemporary landscape.

The nature of both the seas themselves and the sea strand zone around them has also ensured that fish and marine resources can be tapped for consumption in the region and for trade outside it. As this and the following chapter show, the societies of the region developed a range of indigenous skills to tap these important resources, some in harmony with the marine environment, others, especially in recent decades, posing more serious problems of sustainability. Fishery resources today remain rich and diverse but face long-term decline through poor management. Techniques of aqua-culture have emerged to capitalise on the many suitable sites along the coast as well as increasing demand from domestic as well as international markets.

TIN MINING: PAST AND PRESENT

The search for mineral resources (especially tin) in the early colonial period revealed a wealth of information about the geology of the territories on both sides of the Straits. Early in the exploration phase it was clear that there was a link between tin-bearing ores, casiterite and granite and this has resulted in the delineation of major resource areas on the basis of 'granite provinces' or 'granite basement terrane' (Cobbing *et al*, 1986; Gasparon & Varne, 1995; Schwartz *et al*, 1995). Based on their distinctive isotopic and geochemical characteristics, three important provinces have been recognised by Hutchison (1983), despite the range of major and trace element compositions in each one of them. All the three provinces have, to different extents, exerted influences on the development of the Straits, in that the exploitation of tin ores directly impinged on the land, rivers and estuaries of the region. Of the three, the central granite province is the richest in terms of tin ores found and mined to date since many of the rich alluvial deposits of Peninsular Malaysia were found in this province.

The evolution of the Main Range of Peninsular Malaysia saw the intrusion of granite massifs that contained rich tin-bearing rocks. Over geological time these rocks have been eroded, transported and deposited in the river valleys and coastal plains. Unlike the sediments, the heavier tin ores settled before they reached the sea, thus creating the important alluvial tin-bearing areas in river valleys of the Kinta in Perak, and Klang and Langat in Selangor. In addition, the geology of much of the area north of Kuala Lumpur is characterised by large areas of limestone. While blocks of limestone have appeared above ground, a significant proportion is buried underground. Solution processes have resulted in the formation of depressions and pinnacles. It is in the depressions that the heavier tin ores have settled, forming rich pockets of ore. Much of the tin mining activity in Malaysia targeted these rich tin-bearing pockets, but deposits in alluvium nearer the coasts are also widespread. In Sumatra, rich tin deposits are found chiefly on the islands of Bangka, Belitung and Singkep, located to the south of the Straits. The wealth of tin ores in these islands was discovered more than two centuries ago and mining on these islands dates back to at least the early eighteenth century (Donner, 1987, 271). On the island of Singapore, as in the rest of the Riau Archipelago, no rich tin-bearing formations were found, although it is interesting to note that there is a Bukit Timah (Tin Hill) in the island republic.

Large-scale industrial tin mining in Peninsular Malaysia began in the 1870s (see Brookfield *et al*, 1995, 30–32). Prior to that, traditional methods such as panning in rivers were used, but in later years, with the influx of European and Chinese capital, more extensive methods like open-cast mining using hydraulic pumps and palong (paddocks), and expensive dredges were introduced. From these rich fields on the west coast of the peninsula, tin mining grew to be the most important economic activity and export earner of Malaya and, later, Malaysia. Exports of tin concentrates from Malaya rose from 39,569 tons in 1898 to a peak of 72,620 tons in 1972, but declined to 37,874 tons in 1983 (Hew, 1984); this trend has continued since. In her hey-day the country was the largest exporter of tin in the world.

While there is no doubt that tin mining played an important role in the early economic development of Malaya, the methods employed in its extraction were destructive to the natural environment because of extensive land clearance, digging, river pollution and soil erosion. Open-cast mining, gravel-pump mining (which uses hydraulic pumps, water pumps and palongs) and dredging saw the once lush lowland and coastal areas of Perak, Selangor and parts of Negeri Sembilan transformed into landscapes full of scars. Numerous man-made ponds pockmarked the landscape. Forest vegetation was cleared and barren slopes were then bombarded with high-

Table 4.1 Tin mining areas in Malaysia

State	*Area (hectares)*
Perak	5,388
Selangor	13,667
Negeri Sembilan	2,826
Federal Territory	1,809
Kedah	1,803
Johor	19,544

Source: Mines Department Malaysia (1996) *Annual report*.

pressure hydraulic pumps to remove the ore-bearing sediment. Vegetated slopes gave way to highly eroded and gullied landscapes. These features of a degraded mining landscape were vividly portrayed by Ooi in his description of the tin mining landscape of Kinta (Ooi, 1955).

The environmental damage due to mining continues to pose problems. The preparatory work involved in bringing a mine into operation involves the construction of significant infrastructure including kongsi houses (Chinese clan houses), the main buildings and bunds. In addition, the overburden must be stripped before gravel pumps are installed and pump-mining operations begun on the overburden. The extent of damage is appreciable in that gravel-pump mining normally takes place on sites of dredged-out lands for the purpose of recovering tin ore which for various reasons has been left behind from previous mining operations (Hew, 1984, 243). Areas that have been stripped of the ores are then abandoned. These areas are often characterised by mounds of tailings which are occupied by *lalangs* and other naturally regenerated vegetation typical of species that colonise poor sandy soils, as well as ponds which are left totally unused. As Table 4.1 shows, such ex-mining lands in the west coast states of the peninsular are extensive. No rehabilitation measures were adopted as soon as a mine was abandoned, and there was no legislation to make such measures mandatory. It was not until much later, in the 1970s, that some efforts were made to rehabilitate abandoned areas, but these attempts proved too feeble and ineffective to make any impact on land restoration. However, with the rapid rate of economic development in Malaysia in the 1980s and 1990s, and with development land becoming scarce, more attention has now been given to the development of ex-tin mining lands. Capitalising on some of the natural beauty inherent in some ex mining areas, especially the areas of limestone hills and tower karsts, Perak, for example, has given itself the title of the 'state of a thousand pools', a phrase that is designed to attract investment in opening up such pools and ponds for economically useful purposes.

The destructive nature of tin mining using the methods mentioned above has been a cause of concern for many decades. Not only was the natural vegetation cleared and the landscape scarred and modified, but the tailings and slush were often discharged back into the river systems causing constriction of channels and very often flooding.

Furthermore, during heavy rains, the high sediment loads produced find their way into rivers and onwards to the coast. The presence of large areas of depositional landforms along the coast of the straits, and large islands within the confines of broad river channels owes much to the heavy sediment input from such activities. It is not unusual to find that after heavy storms, thick sediment plumes spread outwards from large estuaries into the waters of the Straits. Given the erosive nature of tin mining and the environmental degradation it produced, the colonial government had introduced some regulations to curb the environmental effects of mining activity. However, the effectiveness of such regulations was questionable, given the lack of monitoring and implementation, and the significance of this resource to the colonial economy.

Tin in Indonesia is mainly found in the tin islands mentioned above, whose deposits are similar to the those found in Peninsular Malaysia, Thailand and Burma. The primary ore occurs in veins and veinlets within post-Triassic granites, with cassiterite as the tin mineral. These primary occurrences in turn have been concentrated in many instances in residual hill slope deposits remaining after intense tropical weathering and also as placer deposits in valley bottoms. Quaternary submergence drowned the lower reaches of the river valleys with their tin ores to depths of as much as 30 m. Many such valleys can be traced out to sea and the tin recovered by sea dredges.

As in Peninsular Malaysia, the tin industry on the islands of eastern Sumatra is declining in importance mainly due to resource depletion and price shifts. Because the cost of production in Indonesia is well below the world market price, it is still worth mining lands with a low resource base, which is no longer the case for Malaysia. Outside these tin islands, smaller deposits are found in the Riau Archipelago. Given the substantial fall in the Malaysian and particularly the Indonesian currencies since the mid-1990s, suggestions have been made to open up new mines in less well endowed areas, as well as to reopen some old mines which have been closed but which may still have economic deposits. Whether there will be a revival of the tin mining industry depends very much on other factors as well. These include world market demand that currently ensures net prices remain higher than the cost of production. However, should the currencies appreciate in value, the benefits of opening up of new mines or reopening of old ones would be short-lived. Any reopening of old mines must be considered with great care given the large capital outlay required. Notwithstanding all the above factors, the demand for tin must be on the basis of use, a very important factor that was the cause of the tin market plunge in the 1980s when other, cheaper substitutes for canning were introduced.

The recovery and reclamation of ex-mining land is not only difficult but expensive, depending on the mining methods used earlier in the area. It is also costly when proper account is taken of the sheer extent of land involved. In 1991, the total mining land in Perak alone was 2,100,490 ha as compared with 55,233 ha in 1960 (Lim, 1993). Efforts have been made to develop this land and the abandoned ponds that are the relict features of mining. For example, many ponds have been used for aqua-culture (grass carps and tilapias). The rearing of fresh water fish in such unused pools constitutes an important low-cost production system in Malaysia (Tan, 1992). When combined with the rearing of ducks and pigs, the availability of animal manure and water has also encouraged farmers to incorporate vegetable and fruit farming, which has proved economically profitable. In other areas, the potential of using pools as sources of water supply and for recreation purposes has been recognised. Given greater affluence and a higher standard of living of the population in Perak and Selangor, the need for more water resources, and recreation and leisure opportunities have directed investments into these areas. An excellent example of an ex-mining pool that has been converted to a high class leisure area is the Sungai Way Water Resort, where a hotel cum-recreation complex has been attracting domestic and foreign tourist arrivals. Another is the Mine Mall of Kuala Lumpur, where shopping and recreation complexes and theme parks are combined making use of an old mining pool. However, not all pools can be developed in this way. It is the task of planners to see the potential of each pond and pool taking into consideration their accessibility, size and cost of development.

Where these ex-mining lands are near large towns, the need for housing development has seen such areas being converted for residential purposes. However, the type of houses built is generally single or at most double-storey since the soft foundation of mined land is not suitable for high-rise buildings. The poor quality of soils coupled with issues of pollution on abandoned lands does not provide much potential for farming. Nevertheless, away from towns, growing crops such as guava, groundnut and pomello, which grow quite well on sandy soils supplemented by fertilisers, has produced significant results. The Malaysian Forestry Department has also embarked on reafforestation of these abandoned mining lands, but the choice of tree species is often limited by the nature of the soil. Hence the faster growing and leguminous species like *albizia mangium* and *casuarina equisifitola* have given best results. Other environmental hazards of mining such as the influence of radioactive materials and heavy metals on stream beds and river water have not been well researched. Given the drastic decline of tin-mining activities this industry is truly in its sunset days. Currently, tin mining accounts for only 402 ha of land in the states of Perak, Negeri Sembilan and Johore. Only three dredges, 20 gravel pumps, 13 open-cast and one underground mine are currently in operation in Peninsular Malaysia, and yet in 1982 there were 376 mines in the State of Perak alone (*New Straits Times*, 30 January 1998).

HYDROCARBON RESOURCES

Oil and gas are important resources in Southeast Asia and most are to be found within the broad shelf bordered by the islands of Java, Sumatra and Borneo to the south and the Malaysia–Thailand and Indochinese peninsulas to the north (Siddayao, 1978; Valencia, 1985). The Straits lie within that zone and have important hydrocarbon resources.

It is the western portion of the Straits that has the most extensive worked deposits of hydrocarbons, with the Indonesian provinces of Aceh and Riau being major

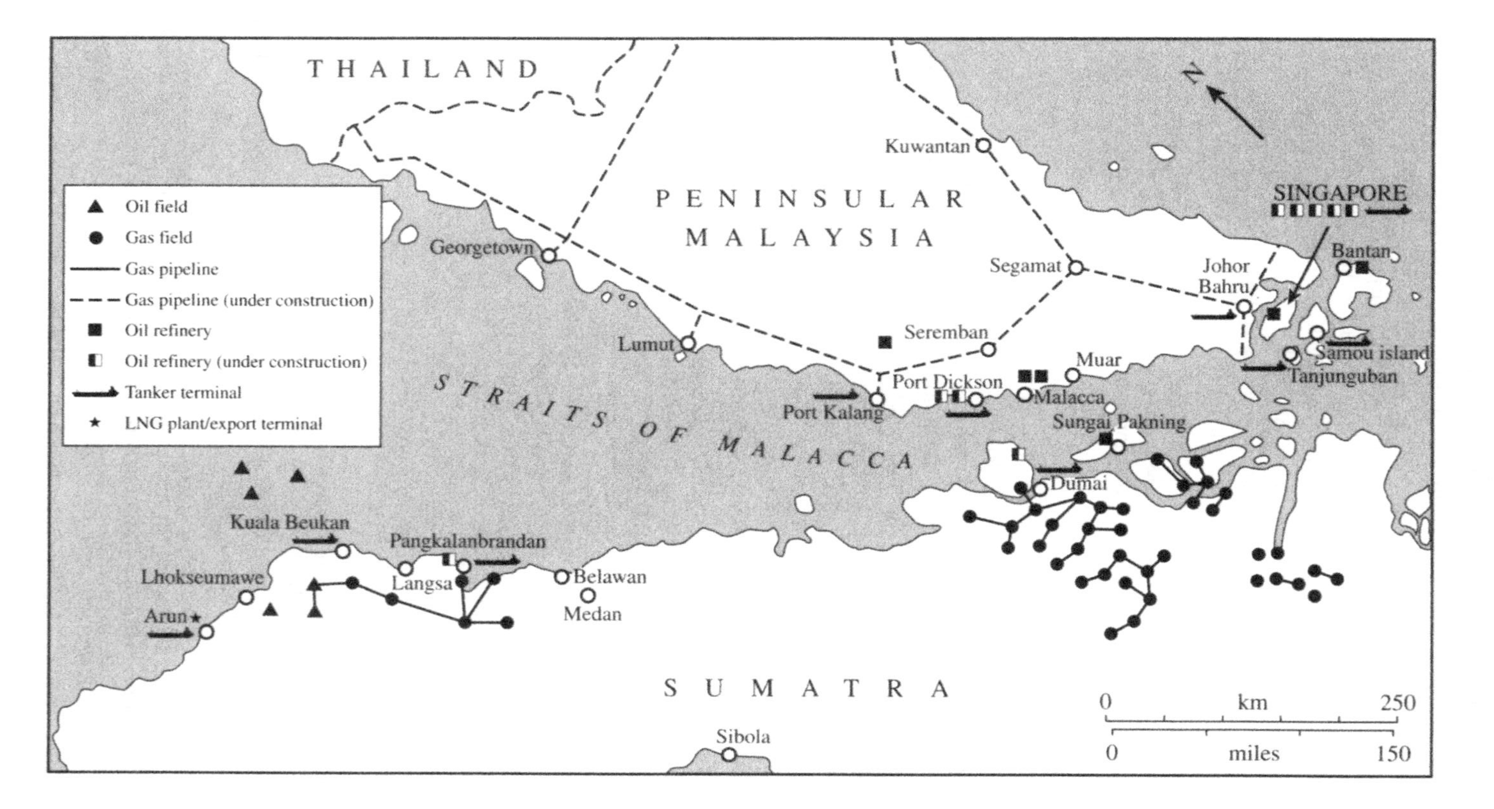

Figure 4.1 Major hydrocarbon production sites and pipelines (adapted from Chua *et al*, 1997, 110).

suppliers of both oil and liquified natural gas (LNG) to domestic and export markets. North Sumatra and the Malaysian states bordering the Straits do not currently have workable fields. The discovery of natural gas in the Lhok Sukon region in Aceh in 1971 led to the development of a production sharing agreement between the Indonesian government and Mobil Oil Indonesia. It was subsequently found that there were large reserves of natural gas in the fields and these were developed to supply six LNG 'trains', five fertiliser plants, an olefin centre, and a number of associated chemical industries (Dayan & Sjafrizal, in Hill, 1989, 115). By 1985, Aceh was Indonesia's third largest source of crude hydrocarbon exports, behind Riau and East Kalimantan.

Petroleum and natural gas are the dominant mining sectors in the Riau Province. The gigantic Minas and Duri fields were the first to be opened, in the 1960s; by the early 1980s a further 52 smaller ones had been developed. The petroleum industry prompted the development of the ports of Dumai and Pekanbaru, as well as improvement in road communication between the two harbours. Total recovery from the Duri field was approximately 450 million barrels, but improved methods of recovery could increase total production to between 2 to 3 billion barrels (Rice, in Hill, 1989, 135). A range of upstream and downstream projects have sprung up as a result of petroleum including the Dumai Hydrocracker Project to produce kerosene, diesel, fuel, petrol, LPG and calcined coke, mainly for domestic consumption.

The overall production of hydrocarbons in the region is declining. Current oil reserves are around 5.5 billion barrels which are proven; a further 5.5 billion potential resource is estimated to be available (Widhyawan Prawiraatmadja, 1997). Given the declining domestic supply in Indonesia and the drive to increase manufactured exports, the role of hydrocarbons is likely to decline. Oil and natural gas accounted for 82% of total export revenue in 1981/82, but their share has fallen continuously since then and is now only 25%.

In terms of production, a peak was registered in 1991 when 1,593,000 barrels/d (kb/d) of crude was produced, but declined steadily from 1,530 kb/d in 1993 to 1,499 in 1995. Indonesia's all-time high was 1,685 kb/d in 1977. Imports of crude oil have grown from 120 kb/d in 1990 and 1991 to 160 kb/d between 1992 and 1994 and almost 190 kb/d in 1995. Indonesia needs about 100 kb/d of Middle Eastern crude to produce asphalt, lubricants and other petroleum products such as light, middle and heavy distillates, since indigenous crudes are ill suited for making these products. It is estimated that total imports of foreign crude will be around 1,300 kb/d in the year 2000. By then Indonesian crude exports are expected to have declined from the current level to about 600 kb/d (Widhyawan Prawiraatmadja, 1997).

ANCILLARY MINERAL DEPOSITS

Important commercial clay deposits are associated with the exploitation of tin in Peninsular Malaysia and they have been used in the manufacture of ceramics, pottery and china, particularly in the state of Perak. Yet in the hey-day of tin mining, much of this valuable clay strata was destroyed. Given the nature of clay, its tiny particle size and colloidability make it easily transported downstream, even at low velocities. The work of the hydraulic pumps, and the washing of earth materials to separate tin ores from stones, sand and clay, made it even easier for clay to be washed away. Thus much of the clay would have found its way to the coasts and been deposited there.

Despite this loss of clay in the past, it is estimated that 40 million tons occur in Changkat Jong, and another 10 million tons in Changkat Cermin in Perak, and 10 million tons of good quality ball clay are exploitable in Dengkil and Batang Berjuntai in Selangor. Aw (1986) evaluated the kaolin resources of Bidor, Perak, which extend over an area of 70 sq km, with more than 50 million tons of deposits. The mining of tin led to the exposure of much of the clay strata. In the mid-1980s, this area was the main clay-producing region in Malaysia and the clay produced was used in a dozen processing plants which produced more than two-thirds of Malaysia's kaolin output of 50,000 tons per annum. Much of the material was used for the ceramic, paper, paint, plastics and rubber industries. The relative abundance of this resource has enabled the State of Perak to set up the Chemor Ceramic Park, complemented by extensive infrastructure to attract investments in ceramic manufacturing. This infrastructure includes air, road and rail links, and a dry or inland port called the Ipoh Cargo Terminal (ICT) to facilitate easy exports.

Limestone geology characterises much of the west coast of Peninsular Malaysia north of Kuala Lumpur. Evidence of marine organisms such as diamictites seem to suggest that these formations share similar origins within a region that extends from Sumatra in the south (Bohorok Formation) through northwest Peninsular Malaysia (Singa Formation), Peninsular Thailand (Kaen Krachan Group), Peninsular Burma (Mergui Group) to northeast Burma (Martaban Series and Lebyin Group) (Metcalfe, 1988). The predominance of limestone in states like Selangor, Perak, Kedah and Perlis has generated cement-making factories whose production caters mainly to the domestic market. As limestone constitutes almost 80% of the raw material requirements, much of this comes from localities close to the factories. In Ipoh, the Tasek Cement Factory, which came into operation in 1964, has now to depend on limestone rocks brought from some distance from the factory because sources nearby have become depleted. While alluvial tin mining digs deep into the ground, limestone extraction results in the scarring of hills towering above ground; while in some it led to the complete demolition of these tower karsts.

FISHING RESOURCES AND MANAGEMENT

The fishery resources of the Straits of Malacca have always supplied the needs of the towns and coastal villages of the west coast of Peninsular Malaysia, the east coast of Sumatra and the urban island republic of Singapore. Historically, the important resources of both the sea and the sea-strand region were vital in both sustaining settlement in the Straits region and in providing opportunities for trade and migration. Today, whilst the fishing industry is not as important as it once was, it still

Table 4.2 Fisheries in Malaysia, 1994

Region	*In-shore landings (metric tons)*	*Value ($)*	*Deep sea landings*	*Value ($)*
West coast	493,268	1,298,719	35,550	71,665
East coast	234,199	458,145	56,447	119,370

Source: Department of Fisheries (1995).

makes a significant contribution to the wellbeing of the coastal communities and the dietary requirements of the populations in these areas.

In Malaysia, marine fisheries can be divided into two sectors, in-shore and deep sea. Of the two, the in-shore sector is declining in importance primarily due to over-exploitation, while there is potential for development in the deep sea sector. The need to increase fishery production for both domestic and export needs has led to the growth of aqua-culture techniques in both fresh water bodies as well as brackish water culture. Of the two, brackish water aqua-culture has become increasingly prominent in the region.

The importance of marine fisheries in the national economy of Malaysia and Singapore has declined over the years, with greater roles today being played by manufacturing and other sectors. In 1988, the fishery sector accounted for only 3.4% of Malaysia's GDP and employed some 2.06% of the national labour force (Majid, 1988). In 1995, the total production from the fisheries sector amounted to 1.25 million tons, valued at RM$3.15 billion, accounting for some 1.47% of national GDP (Department of Fisheries, 1995, 42). These figures apply to Malaysia as a whole, but the Straits region and the coastal areas of Peninsular Malaysia bordering it are more significant than the east coast, in terms of total fish catch as well as the number of people directly or indirectly involved with fishing (Table 4.2). There are important reasons for this difference between the west coast and the east coast of the peninsular. The Straits of Malacca have calmer waters and fishing can be carried out throughout the year, while the seas off the east coast of the peninsular are subject to strong northeast monsoons. The strong winds and high waves during the period of the northeast monsoon make it hazardous to go out into the open sea to fish, whereas in the Straits these seasonal impacts are much less marked. Another factor is that traditionally, the Straits region was more densely peopled, with greater demand for fish products. Consequently, the Straits of Malacca are the most intensely fished area in Malaysia, and it is the Straits region that has the best developed fishery industry. Potential catches are high in these areas, as Figure 4.2 suggests.

Despite the general decline in the 1980s, over the past few years total catch has begun to increase. The total annual landing in the west coast of Peninsular Malaysia in 1995, for example, showed an increase of 14.89% from that of the previous year. All the states in the west coast of the peninsula except Negri Sembilan showed an increase in their total marine landing (Department of Fisheries, 1995, 46). As far as in-shore marine landings are concerned, the west coast of Peninsular Malaysia accounted for about one-half (493,049 tons) of total landing in the country in 1995.

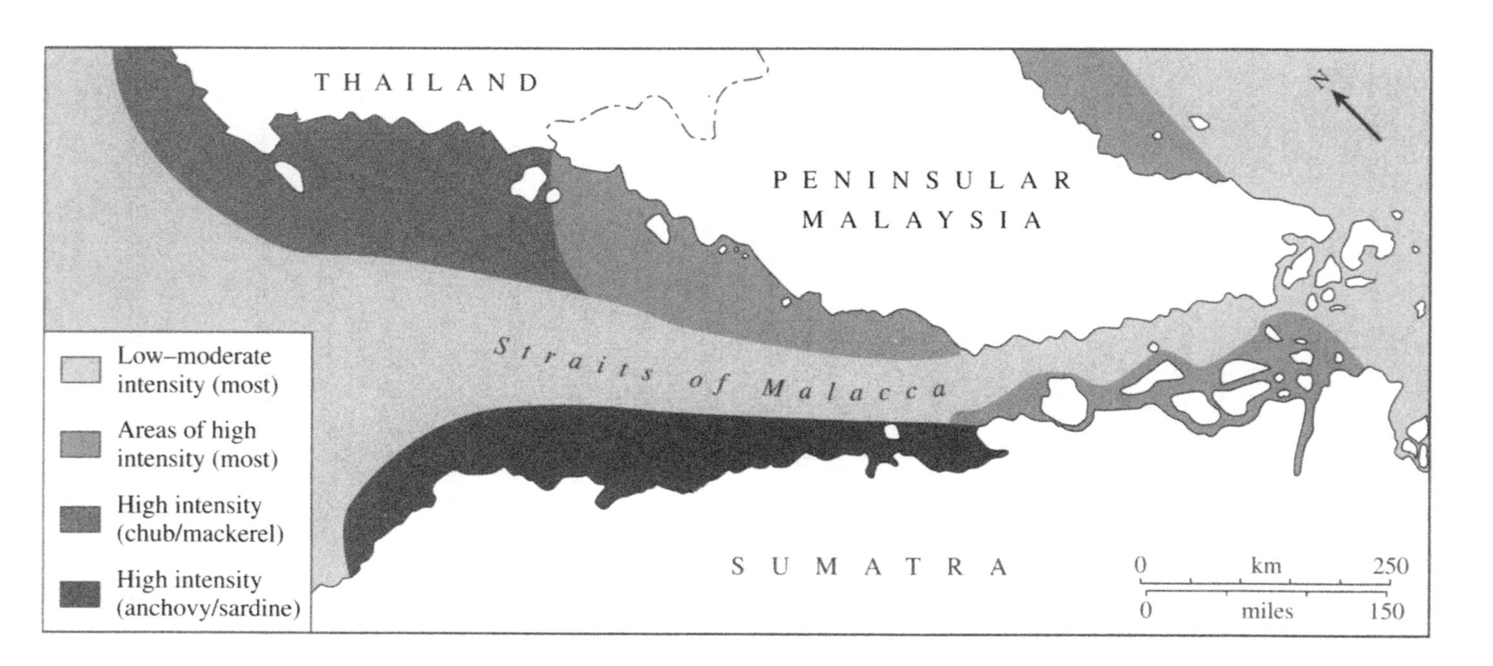

Figure 4.2 Productivity of fishing grounds in the Straits region (adapted from Valencia, 1985).

By comparison, the east coast accounted for just over 24%, with the rest being landings from Sabah, Sarawak and Labuan. Despite overfishing, the west coast still showed steady increases in landing in recent years; it was the east coast that showed a decline. Given the greater area of sea for deep sea fishing, the east coast produced some 40.72% (56,447 tons), while the Straits contributed 25.65% (35,550 tons).

One of the problems contributing to fluctuations in the catch in the Straits has been the lack of regulation concerning the type of gear used and hence the size of fish caught. Arguably, the introduction of modern trawling techniques has had the greatest impact on the fish resources of the Straits (Menasveta, 1994; Marr, 1981; Yap, 1977). Traditional methods of fishing in Malaysia prior to the 1970s were essentially artisanal in character, producing a catch that was only sufficient for domestic consumption. Techniques used included stake traps, beach seines and cotton grill nets and the total catch in 1959, for example, was less than one-quarter of the catch 20 years later. After the introduction of trawler fishing in 1970, total catch for that year increased threefold to 338,500 metric tons, reaching its peak in 1980 with 733,200 metric tons. This period of expansion coincided with an increase in the number of trawler boats, expansion of the commercial fishing fleet and the construction of fish-landing facilities such as cold storage and ice plants. This trend of growth in the catch did not last and by the mid-1980s, what had appeared to be a good investment in trawler fishing seemed to have collapsed. Total catch began to decline and there was an increase in illegal trawling. One consequence of this was an increase in the landing of trash fish. Before 1966, less than 16% of Malaysia's total catch were trash fish. By the 1970s, the proportion had increased to 33%. Trawling accounted for some 70% of trash fish (Huat, 1978).

A large portion of trash fish consists of juvenile fish, which if allowed to mature would constitute an economically important catch. In Selangor, on the west coast of the peninsula, for example, there was a 20% decrease in edible fish landing in the mid-1970s, coupled with a doubling of landings of baby trash fish, indicating that the matured fish stock was being eroded (Yap, 1977). Modern methods of trawling, combined with new gear that has made fishing more effective, proved to be a recipe for over-exploitation in the Malaysian waters, particularly within the Straits. Ironically, the Malaysian government announced in early 1984 that the National Agricultural Policy was to emphasise the continuous use of modern methods to fully exploit the fishery resource and to increase off-shore fishing (Majid, 1988). Evidently, at that time, it appeared that the authorities failed to realise that the introduction of trawler fishing and the lack of control on their activities or number, legal or otherwise, was the prime contributor to the decline in the total fish catch (Marr, 1981).

The onset of more efficient methods of fishing such as trawling with Apollo drag weight nets and purse seines also saw conflicts between those fishing which sometimes ended in loss of boats and lives. The waters of the Straits, especially the extensive mud banks, are an excellent fishing ground for in-shore fishermen. Trawling was designed strictly for off-shore fishing, but the desire for quick profits from higher valued fish has led to frequent encroachments of trawlers into the shallower coastal waters. This problem continues to affect local fishermen. Recent problems off the waters of Johor, at the southern end of the Straits, exemplify the difficulties. More than 100 coastal fishermen met at Bagan near Batu Pahat to protest against illegal trawling activities (*New Straits Times*, 20 November 1997). Six trawlers were detained by the Fisheries Department on 21 February 1998 for encroaching into coastal fishing grounds. They were caught some seven nautical miles off the Bagan coast. More than 1,000 coastal fishermen have been constantly harrassed by illegal trawling, which has damaged fish breeding grounds and destroyed the fish traps and nets of inshore fishermen (*New Sunday Times*, 22 February 1998).

The nature of the trawling process has also contributed to stock diminution, especially in shallower coastal waters. Thus, nets can scrape the mud bank and shallow sea-bed, taking in their path all sizes of fish, prawns, baby cockles and crabs, and juvenile fish and damaging the sea-floor. The ecological damage to the shallow waters has meant disruption to the activities of in-shore fishermen, as well as the need to import fish when once there was sufficient supply from these waters. Merican (1977) has noted that at the height of trawling activity overall fish exports from Peninsular Malaysia actually fell and imports rose. The negative balance of trade in Malaysian fishery products is clear evidence of over-fishing and over-capitalising through the indiscriminate use of trawling as new fishing gear. Overfishing has resulted in disruptions to fish growth cycles and has also compromised the checks and balances of the food chain through changes to the species ratio (Frankel, 1995).

These difficulties have been compounded by illegal fishing and by the use of destructive fishing methods. These have included the use of dynamites, poison and sometimes electricity, even though these were in direct violation of marine regulations. There have also been reports of fishermen using tuba roots and pesticides to poison fish (Huat, 1978). Such methods are destructive of the biological cycle of fish, causing the supply of certain species to break down. Such problems may pose special difficulties for the important shellfish industry. Certain parts of the Straits in recent years have become important grounds for cockle and crab production. Tidal flats and calm shallow seas are natural food-producing areas that abound on both sides of the Straits. The tropical cockle, for example, is particularly abundant in the mud flats bordering the water body. It serves as a lucrative form of marine resource cultivation in Malaysia. In specific areas of the coast, shrimps for the production of *chincalok* and shrimp paste are caught seasonally. This is particularly important in coastal areas off Penang and Malacca, between the Tanjong Keling and Klebang Kechil.

Fishing and pollutants

Changes in the fishing catch are revealing, not only of technical changes in fishing methods, but also of the problems of pollution in the Straits. As the second-busiest sea-lane in the world, they experience massive churning of their waters, and the receipt of oily discharges sometimes from tanker collision and more often from desludging, and deliberate dumping of waste, all contribute to the deterioration of their environment. It has been estimated that about 1,000 to 3,000 gallons of oil residue is discharged into the sea with tank wash water on a single voyage of a 200,000 ton tanker plying the Straits. Large concentrations of such ballast discharge have been discovered at both ends of the Straits (Valencia, 1991, 279–290). Apart from direct oily discharges, collisions of vessels can also result in oil spillage and the detergents used to clear these may have deleterious effects on marine life. In the event of oil spills sinking to the bottom, or spreading over a large surface area, the effect on fishery resources and their breeding grounds would be disastrous. Such incidents have caused one commentator to describe the Straits of Malacca as one of the world's 'dirtiest maritime back lanes'.

It is apparent that there is a dilemma between keeping the Straits as a sea-lane of enormous economic importance and maintaining it in an ecologically sound condition so as to be a source of fishery resources and healthy marine life. There is evidently also an inherent conflict between the interests of the private sector that sees the Straits in terms of their primary function of commerce, trade and shipping, and the government agencies that wish to see healthy and clean waters.

While oil spills from ship-based sources have their distinct impact on the marine life of the Straits, indiscriminate dumping and discharges of effluents from factories and industries and wastes from residential areas into rivers constitute a further source of pollution. These land-based pollutants will reach the estuaries and shores first, places where the in-shore subsistence fishermen depend for their livelihood. Lessons from the early 1970s in Peninsular Malaysia have shown the extent of the damage such land-based pollution could cause. The Juru River in Seberang Perai, Penang, for example, became dead from the toxic wastes discharged by factories from the Perai Industrial Estate, located within its catchment, threatening the livelihood of 300 coastal villagers at Kuala Juru.

Besides pollutants from factories, the use of chemical fertilisers and pesticides in paddy planting has not only affected the fresh water fish life in the paddy fields, but also fish production along the shore. A study by the Ministry of Agriculture and Fisheries indicated that there was a fivefold decrease in the production of fish from the Krian District of Perak alone in the 1970s (Idris, 1977). In Penang, a study on the concentration of *coliform* and *e. coliform* bacteria, associated with human and animal wastes, found that the microbial levels in the waters off the northern coast of Penang where concentrations of fishery resources exist, were extremely high (Tan, 1992). The consumption of such fish from these waters may pose a health hazard to local communities.

Apart from these consequences of land- and sea-based pollution that affect the ecological health of the fishing ground and, ultimately, the fish catch, the Southeast Asian seas have been known to have experienced the red tide phenomenon. Past incidents of this phenomenon have been recorded in the South China Sea, for example the coasts off Sabah and Brunei, and much less in the Straits. The causes of this highly toxic algal bloom are still unclear though there seems to be a coincidence of the bloom with rich organic discharge from estuaries consequent on heavy rainfall. However, not all heavy rainfalls causing heavy discharges into the sea bring about red tide phenomena. The problem with red tides is that fish or marine life caught in such areas contains these toxic substances which have fatal effects on humans who consume them.

Developments in fishery management

The fisheries industry in Malaysia and Indonesia has seen some tightening of laws and greater environmental concern in recent years. It was evident in the Straits that the rapid increase in fish catch in the 1970s, without proper assessment and management, made the industry economically and ecologically unsustainable. Consistent with environmental management efforts in other areas, the Malaysian government in particular has made a concerted effort to tackle the problem of declining fish stock in the Straits. With this drive came the formulation of a range of rules, regulations and objectives for the control of fisheries, as well as their stringent enforcement, with priority given to ensuring the sustainability of fish stock well into the next century. A comprehensive fishery licensing policy to regulate commercial trawler fishing was introduced in 1981. Since then there has been a moderate decrease in the non-selective and destructive impact of fisheries from trawling. As a result of these series of measures, the last decade has seen a partial revival of Malaysian fisheries as indicated by its increased catch.

By the same token, the revised Malaysian Fisheries Act has made further provisions for the conservation and protection of fisheries in estuarine and marine waters and for the exploitation of fish in these waters by citizens or Malaysian fishing vessels. For example, the turtle excluder device (TED), adopted in 1997, is one of a number of measures introduced by the Malaysian Fisheries Department to conserve and protect certain endangered species. The TED is a device made of steel or aluminium rods inserted into fishing nets. It is designed to prevent turtles from being trapped in the nets but does not affect the average catch of fish and prawns (*The Straits Times*, 28 February 1997). Much earlier, as another example, the Malaysian government banned British Airways Concorde from flying over the Straits of Malacca to prevent sonic boom from disturbing spawning fish (Valencia, 1991). The Act also includes conditions restricting fishing to specified zones and times, prescribing minimum size limitations on mesh size used in fishing gear, and heavier penalties for breaching regulations (Menasveta, 1994). Enforcement of the Act is undertaken jointly by the enforcement section of the Fishery Division and the Marine Police. The effectiveness of this enforcement has been heightened by the increase in the number of coastal boats and frequency of patrol to deter illegal fishing and over-exploitation of fishery resources (Majid, 1988).

As other economic activities have developed, fishing has been of only minor significance in Singapore. Much of its supply of sea fish comes from neighbouring countries. Yet in the 1940s and 1950s, more than 40,000 tons of marine fish were caught annually. By 1990, the number of fishermen registered had declined to just over 2,000, as had the number of *kelongs* (palisade traps) and motorised vessels. The availability of alternative employment opportunities was a factor in the decline in sea fishing in Singapore, but it was not the only factor. Another important element was the loss of coastal fishing grounds due to extensive reclamation of coastal areas from the 1960s, removal of coastal mangroves, and the extensive canalisation of rivers and streams that reduced the amount of sediments and nutrients flowing into the sea. Reclamation along the coasts also resulted in increased sedimentation of coastal waters. The development of much more extensive port facilities meant increased shipping activities and increased inputs of oil and other pollutants. To protect coastal waters for other important uses, fishing licences were restricted.

Fishing and fish resources in the Straits are of major importance to the eastern coastal region of Sumatra, except that the population pressure along this coast is not as great as that of the west coast of Peninsular Malaysia. Nevertheless, fishing has traditionally been an important activity of coastal villages from north to south Sumatra along the Straits of Malacca. However, in proportion to the total fish catch of the whole of Indonesia, the contribution from the Straits is not especially significant. Similar problems have beset fishing in the Indonesian waters of the Straits to those in Malaysia. By the late 1970s, the major Indonesian fishing grounds were thought to be fully exploited or overfished, particularly the demersal fish and penaeid shrimp which were considered to be fully exploited in the Malacca Straits, as well as in other areas of Indonesia.

The problems posed by trawling also affected the Indonesian waters of the Straits in the 1970s. Frequent conflicts between fishing trawlers and artisanal fishermen in seas which are already overfished caused the government to ban the use of trawlers in western Indonesia under Presidential Decree No. 39 of 1980 (Rice, 1991, 158). This ban was extended to all of Indonesia in 1983. Initially, the ban caused a drop in maritime shrimp production, but that trend has recovered gradually through the use of other types of fishing methods and boats. The most beneficial effect of the ban was the increased catch by the thousands of artisanal fishermen. Certainly the effect of the ban caused an increase in fish catch in the provinces of Sumatra bordering the Straits. Riau province showed the smallest increase in volume of fish production between 1979 and 1987, yet the number of fishermen increased sharply, obviously affecting the overall productivity per fisherman. Aceh and North Sumatra provinces benefited from the new regulations. But fishing resources remain scarce and in demand: price surveys suggest that fish prices in Medan over the last two decades have been rising at a consistently higher rate than the consumer price index. Apart from over-fishing in the Sumatran side of the Straits, there is very little data on pollution to support the contention that the coastal waters there are polluted. Certainly there is evidence of high sediment discharge from rivers that flow into the Straits from the three provinces of Aceh, North Sumatra and Riau, but quantitative data is unfortunately lacking.

International cooperation in fisheries protection

There has been a greater awareness of the need to develop and manage the fishery resources in the Straits of Malacca by the states bordering these water bodies in the last few decades. While it is recognised that the Straits are waters that have been overfished, there is an awareness that cooperative action can help maintain sustainable catch levels. A proper strategy and policy towards the development and management of fishery resources in these waters must be developed and seriously implemented by the countries themselves in so far as their own territorial waters are concerned. However, where the waters are regarded as common grounds, cooperation among the three littoral states is absolutely necessary. However, the development of fishery resources must of necessity extend to the international level, and this has been and will continue to be addressed by the three countries. Cooperation among the three countries is within the framework of ASEAN at the regional level. The ASEAN/US Coastal Resources Management Project is one good example of cooperation in managing the resources in the coastal zone. Under this pilot project, studies have been carried out to provide baseline data for the coastal zones in all the ASEAN countries. Through these a number of strategies and actions for coastal zone resources management have been proposed for each country, which also include the coasts bordering the Straits of Malacca.

In the past, Malaysia, like Indonesia, has tried to solve its fishery resource problems from the perspective of its own needs and requirements through its own ministry. However, by coordinating initiatives with other ASEAN members, it has been possible to reverse the negative effects of depleting fishery resources in the Straits. The interests ofThailand must also be included, for far too often in the past incursions into the territorial fishing areas of Malaysia have been made by Thai fishermen, giving rise to unpleasant incidents. Thus, at the Third ASEAN Ministerial Meeting on the Environment, held in Jakarta in 1987, a statement was made to the effect that, 'ASEAN member countries adopt the principle of sustainable development to guide and to serve as an integrating factor in their common effort' (Leekpai, 1991). The participation of Malaysia provides for international actions to agree on management objectives and sustainable development in coastal seas that are important for the fisheries and at risk from pollution in the 1990s. (Holdgate, 1992). At the international level, Indonesia and Malaysia are signatories of the Fish Stocks agreement arising from the United Nations Conference on Straddling Stocks and Highly Migratory Fish Stock Agreement in 1995. The agreement requires that special efforts be made to monitor fish stocks. This agreement will be an official supplementary document to the UN Convention on the Law of the Sea, which nearly all fishing nations abide by (World Resources Institute, 1996).

AQUA-CULTURE

Throughout Southeast Asia, aqua-culture has become an important activity that has resulted in the conversion of mangrove swamps and coastal areas into fish and shrimp ponds. This trend is growing partly because, though capital intensive in the initial stages, the returns are more secure and the supply can be easily adjusted to meet the demand from both domestic and international markets.Within the coastal areas of the Straits aqua-culture activities are evident.

In Sumatra, *tambak* or brackish water ponds, are used to raise milkfish and shrimps. Four methods of farming have been identified by Rice (1991, 167)—traditional extensive, improved extensive, semi-intensive and intensive. The first method depends on high tides bringing brackish water to into the ponds that supply the nutrients for the milkfish and shrimps commonly raised together. In the second method, while tides are required to bring in water and nutrients, farmers also supplement this by using fertilisers to encourage algae growth. The standard semi-intensive system uses the higher density of shrimp method (27,500–42,500 post-lava 30 fry per crop per hectare). Prepared feed is added to the ponds and pumps are used to increase the rate of water exchange. The last method adopts a much higher shrimp density, which means that the water needs to be changed more frequently and the dissolved oxygen content in the water kept at a healthy level by using the mechanical paddle-wheel aerators. In this method, pure sea water is pumped into the ponds simultaneously with fresh water at a predetermined ratio in order to obtain the desired salinity level. In addition, the shrimps depend entirely on manufactured shrimp feed given at regular intervals. Apart from ensuring a greater supply of fish and shrimps, the development of brackish-water aqua-culture is regarded as an important avenue for creating productive employment opportunities for fishermen, as well as a means of generating income and foreign exchange, particularly through the export of high-value shrimps. While data on the number of fishermen involved in aqua-culture and total production for the whole of Indonesia are available, statistics for the Sumatran side of the Straits are lacking. Certainly, by virtue of the extensive peat swamp areas in the eastern coasts of Sumatra, the potential for increasing the *tambak* area in this region is great. There is also potential for intensification of the main *tambak* areas of Aceh and North Sumatra to complement efforts made in Kalimantan and other parts of Indonesia.

Indonesia has greater advantages in aqua-culture than her neighbours bordering the Straits. Extensive swamp areas can be opened up for this activity, and labour, at the different stages of aqua-culture development and operation from pond construction to the processing of shrimps for export, is plentiful and cheap. However, there are also several constraints, including the high costs of manufactured shrimp feed, lack of expertise in the production methods by farmers and extension agents, transport and communication costs due to the long journeys and hence the need for good refrigerated containers which are in short supply. Other problems faced are environmental. Sedimentation, pollution of the water through fish waste and feed, and the use of chemicals all can affect output.

The problems over fish catches in the west coast of Peninsular Malaysia have diverted some attention and investment to aqua-culture. Traditionally, aqua-culture activities have mainly concentrated on the growth of cockles for the domestic market. Cockle beds are found in estuaries and off the coast where mud predominates and they are grown with little input of fertiliser other than from organic discharges from land. Over the past two decades aqua-culture has assumed an important role in complementing sea fishing. Several important areas of fish and prawn culture have been opened up from Perlis to Johore and the livelihood of those who are connected with it has improved considerably. By 1985, two important projects were the Sungai Merbok in Kedah and Sungai Danga in Johore. The former had 70 ponds, with a total pond area of 26.4 ha, while the latter comprised 17 ponds, with a pond area of 13.5 ha. In these early days of aqua-culture development, the Fisheries Development Authority (FDA) was instrumental in developing the above two areas. It was subsequently privatised in line with the government's privatisation policy. The FDA has also implemented other schemes involving cage culture of fin-fish, cockle culture, hatchery and supply of fry and technical assistance (Khalil, 1985).

In 1995, some 18,466 fish farmers/culturalists were engaged in the aqua-culture business in west coast Malaysia. In that same year the total production of brackish water culture was 114,000 tons, up from 95,000 tons the previous year. The increase in production came from cockle culture, brackish pond culture and oyster culture which more than offset the decline in fish cage culture and mussel culture. Cockle culture accounted for some 87% of total production of brackish water culture, with about 52% of the total by value. The states of Perak, Penang and Selangor were the main contributors. Overall, brackish water culture showed an increase of 9.84% in value over the previous year, with a total value of RM262.67 million.

Such aqua-culture activities are beginning to occupy larger and larger areas of the coast. In 1995, brackish pond culture occupied an area of 2,623 ha, brackish cage culture 715,152 sq m, cockle culture 4,752 ha, mussel (*Perna viridis*) culture 62,029 sq m and oyster culture 83,317 sq m (Department of Fisheries, 1995, 52). As far as the culture of blood clam (*Anadara granosa*) is concerned, its breeding depends on natural spats for its seed stock. In 1989, the spatfall areas occupied some 4,000–5,000 ha of mud flats in the west coast of Peninsular Malaysia. Green mussel, on the other hand, requires rafts with hanging ropes of jute or polyethylene for spat settlement. The locations of such culture grounds for both clams and mussels make them vulnerable to pollution, an increasing problem to contend with in the Straits of Johor where mussel is still dominant, and the coastal areas of the Straits of Malacca where cockles are widely farmed. Tiger prawns (*Penaeus monodon*) form an important and economically valuable species reared in brackish water ponds; they account for some 90% of total production. Other species include banana prawn (*Penaeus merguiensis*), sea bass (*Lates calcarifer*), grouper (*Epinephalus tauvina*), mud crab (*Scylla serrata*) and other miscellaneous species. The importance of the tiger prawn lies in the high demand from restaurants and supermarkets at home and abroad.

In general, then, there has been an increase in the production of brackish water culture in all the west coast states of the peninsula with the exception of Kedah and Perlis. Attempts have been made to increase production in the state of Kedah through land acquisition along the coast which, due to the manner of its proposed acquisition, caused so much controversy and protest among village dwellers that it has been shelved indefinitely by the new state government. It is increasingly recognised that there is considerable potential for aqua-culture development and production in the Straits region. Together with off-shore fisheries, aqua-culture development has been given a high priority by the Malaysian government (Ong, n.d., 1). It was estimated in 1985 that annual production in the year 2000 would be more than 200,000 tonnes

(Tengku Ubaidullah, 1985), a figure that is not impossible to attain going by the 1995 production figures.

While there is great potential for further development of coastal aqua-culture, the ecological constraints to successful culture activities must be recognised. Different species require different sets of ecological conditions dictated by their peculiar biological requirements. A very important criterion for a suitable site is one which is free from pollution. Many such suitable sites are presently occupied by mangrove swamps and wet lands, and the need to protect such areas for the perpetuation of living resources and as valuable breeding grounds for marine fishes and other organisms means that any coastal land alienation for aqua-culture must be done with great care. Extensive mud flats on the west coast are still available and suitable for cockle farming and some 100,000 ha of coastal or brackish water areas are suitable for brackish water pond development, especially for tiger prawn culture. Any large scale opening up of coastal land for aqua-culture (especially tiger prawn culture) must be done judiciously, taking into consideration environmental and social concerns, the former, issues like acid runoff and high iron content in brackish water, and the latter, conflict between landowners and private or public enterprises.

In addition, the large scale conversion of coastal swamp lands for quick profits should be curtailed by increasing productivity. Experiments in Gelang Patah in Johor have shown that yields of prawns can be as high as 8,000 tons per ha per crop harvested after four months of culture, and in some other sites much higher than this. Current average yields are much lower at about 2,000 tons per year. The above factors, together with government support through research by the Fisheries Department (Ong, 1985, 9–16), extension services (Tan, 1985, 17–21), and success in semi-intensive and intensive grow-out methods and hatchery technology for large scale prawn farming (Seow, 1985), will see the expansion of brackish water aqua-culture in the years ahead. And the coastal areas of the Straits will play a significant part in this.

In Singapore, the bulk of the fish and prawns consumed has to be imported from neighbouring countries. However, attempts have been made to increase domestic production of fish through aqua-culture activities. Given the shortages of land, any alienation of coastal areas for this purpose will not be cost effective, and so such areas will have to be within the water itself and in localities where they do not interfere with shipping and other major economic purposes. Two main areas have been designated for aqua-culture and they are on the eastern and western parts and within the Singapore side of the Straits of Johor. While production has increased, the total fish production accounts for only a small percentage of domestic demand (Cheong, 1990). One of the problems encountered in aqua-culture in this area is water pollution within this Strait of Johor, partly because of the causeway across the strait linking Johor to Singapore, which cuts the flow of water through the channel, and partly due to pollution from the rivers from Johor such as the Johor River, Tampoi and Skudai. Despite these difficulties, these are the only areas considered suitable for aqua-culture. The waters in the south of the island cannot be used for aqua-culture as it will interfere with ports and recreation.

The coastal areas of the Straits of Malacca are thus an increasingly important location for aqua-culture development, where different species have been introduced and various methods for their production have been implemented. The importance of these coastal areas can only grow over the coming years. Yet with intensification of these activities, there are serious environmental implications that need to be addressed at the same time as further expansion and development progress. Modern and intensive aqua-culture systems introduce inputs of water, feeds, fertilisers and chemicals and invariably more wastes into the coastal and marine environment. Consequently, problems of nutrient and organic enrichment, toxicity to the marine environment, antibiotic resistance in pathogens, effects on biodiversity, human health hazards, habitat destruction and sediment accumulation are issues of real concern (Brzeski & Newkirk, 1997). In view of the above impacts, the sustainability of modern shrimp farming practices is increasingly being questioned (Larsson *et al*, 1994; Phillips *et al*, 1993).

The Straits of Malacca, then, constitute a region which is rich in a range of resources, the presence of which has had an important impact on the nature, scale and geography of development in the region. Natural resources from the land—tin, stones, clay and hydrocarbons—combined with the wealth of the seas, has led to the establishment of important industries in the region. The expansion and contraction of those industries have also led to reappraisals of resources, to their reuse in some cases and, in others, to a range of environmental problems with which the governments of the coastal states are having to grapple. It has been the combination of human ingenuity and the resources of the environment that has shaped the development of the Straits region both in the past and present. Resources and technology form an indissoluble whole; it is to the role of technology that the book now turns.

5
TECHNOLOGY, RESOURCES AND DEVELOPMENT

The relationship between the physical resources and cultural use of the Straits has been an important and dynamic element in the historical and contemporary geography of the region. As the preceding two chapters have shown, the geology, morphology, vegetation and soils of the Straits region have provided an enormous range of opportunities for human exploitation. Minerals, hydrocarbons, the fishing industry and aqua-culture exemplify contemporary patterns of resource exploitation in the region. In the past, precious metals and jungle products were part of the currency of trade. It is clear, then, that the ways in which the Straits, their currents, coasts, inlets, mangroves and sheltering places have been used reflect a number of different factors. Technology, in particular, has always been of key importance—how has human ingenuity and innovation in fields ranging from boatbuilding and navigation to corporate organisation and the enforcement of monopolies changed the ways in which the Straits have been perceived, appraised and used? This chapter examines these relationships and seeks to show how the resources of the Straits region have been employed to develop the cultures and societies of the area. This is not to suggest a form of crude technological determinism—the immense varieties of development in the Straits refutes any such arguments—nor, indeed, to suggest any simple evolutionary model of change in the region. Change and development have been immensely complex and that complexity is illustrated here through an examination of historical patterns of exploitation in the Straits.

SEAS, CURRENTS AND NAVIGATION

The monsoon wind patterns are the fundamental and best-known factor which has structured patterns of long- and short-distance sailing through and in the Straits. As we showed in Chapter 2, these patterns have had an important bearing on the destinies of the key ports of the region (Dobby, 1950, 31–46). The year can be divided into two monsoonal periods (see Figure 2.5). The northeast monsoon blows from about December until early February and on these southward winds sail ships voyaging from China or India and the Gulf can reach the Straits. The southwest monsoon, which blows from mid-June to mid-October, then acts to take shipping from the Straits to either China or India on its northward winds. During the transitional period, especially in the months of October and April, sailing ships had to wait for the monsoon to 'turn' and the ports of the Straits, located in what were called these 'lands below the winds', provided suitable anchorages for long-distance ships in the Straits in the pre-steam era. Around these anchorages sprang up the wealth of entrepot and trading functions that have long animated the commercial life of the Straits.

Local patterns of land and sea breezes have also impacted on navigation and shipping. These breezes can be experienced at anything from 25 km to 80 km off-shore and reflect the differential rates of heating of land and sea masses. Cooler air blows on-shore during the warm daytime period, with the pattern reversing late at night and in the early morning. These breezes, furthermore, differ from one end of the Straits to another—weaker at the narrow, southern end, they strengthen further north where the Sumatran and Malayan coasts are as much as 200 km apart.

Whilst the Straits are generally free of strong storms and are well away from the main tropical typhoon zones, rough seas in the monsoon period habitually made navigation difficult. At the northern end, the southwest monsoon blowing past the northern tip of Sumatra could create sailing difficulties. In a following wind during the monsoon, small boats running before the wind faced possible swamping hazards, whilst an on-shore monsoon could raise heavy surface swells making sailing into river mouths and harbours difficult. As Sopher (1987, 29–30) comments, however, ' . . . much of the coastal area, . . . is characterised by weather suitable for quite small and simple sea-going craft during 8 to 10 months of the year . . . certain almost landlocked parts of the sea are suitable for primitive sea-farers throughout the year'. Conditions in much of the Straits, then, have long lent themselves to communication, trade and interaction by sea.

The nature and shape of the coast, when coupled with prevailing weather conditions, have also exercised an important control on the shipping patterns of the region. In earlier times, navigation depended on a process of hugging the coast, coasting, and the availability of suitable inlets for ports was very important in facilitating longer journeys. Much of the coastal zone of the Straits is flat with extensive mangrove formations flanking both coasts. On the Sumatran side, this featureless, seemingly impenetrable zone, is especially extensive with long, winding rivers such as the Siak, Kampar and Rokan fringed by many miles of nipa palm. Silt deposits often clog up these estuarine regions and create a landscape of mangrove, mud flats and expanses of shallow brackish water. This complex network of flats, sand-bars and islands required, and continues to require, all the navigational skills of indigenous sailors.

Rocks and cliffs are less frequent in the Straits than elsewhere in the region. They are noteworthy and pose problems only in a few places. The Langkawi group at the northern end of the Strait, rises steeply out of the sea whilst Pulau Pinang, the Dindings, the granite and quartz outcrops of Singapore island and some of the hills of the Riau group of islands are the other main cliff areas. But, generally, they provide few hindrances to navigation and anchorage.

The nature of the coastal terrain has had a number of consequences for the establishment of ports. As a rule, few prime sites immediately suggest themselves for ports and those that were established owe more to human ingenuity and effort than natural advantage. There is no geographical determinism in their location. Thus, on the Sumatran side in particular, suitable port sites were often dozens of kilometres inland. Palembang, for example, the key port of the ancient kingdom of Srivijaya in the fifth to eighth centuries, was located some four days' sailing from the Straits themselves. But certain sites did have particular advantages—Malacca and Singapore suggest themselves, or Aceh at the northern tip of Sumatra. Generally, however, the rise (and fall) of ports was a function not of site or geographical advantage, but rather of the ebb and flow of trade and human endeavour. As political allegiance, trading endeavour and the growth of the hinterland shifted, ports were quickly created and abandoned.

SHIPPING TECHNOLOGY

There are important relationships between shipping technology and the use of the resources of the Straits but that relationship is not simple. The development of new systems of shipbuilding, navigation or propulsion inevitably changed some of the ways in which the Straits were used, but it is important to emphasise that, alongside the technological improvements, older ways co-existed. Thus, when steam ships first plied the region in the late nineteenth century, sailing ships of all shapes and sizes continued to trade and to use the traditional resources of the Straits. Technological change did not produce a simple, linear evolution in the ways in which the seaways in the region were used.

The *prau* (or *prahu* or *balok*) was the basic form of sea transport in the Straits and its design and function dominated short-distance indigenous trading. Alongside the *prau* was the junk which was the main longer-distance trading vessel used in the Straits for some 1,500 years at least. The word junk came into the English language, not from the Chinese, but from the Malay term for this characteristic boat, the *jong*. Its development and form seems to represent an amalgam of Malay and Chinese shipbuilding techniques (Reid, 1993, 36). Evidence that junks sailed in the Straits date from at least the fourth century AD and, unlike western ships, its interior was divided into a series of compartments (Kemp, 1980, 43–52). The junk shared many of the characteristic features of the *prau*. It had a keel to which planks were joined, often with wooden dowels, and was usually equipped with a double rudder. Unlike the *prau*, which usually had a single sail, the typical junk had two or more sails as well as an upturned prow (Figure 5.1). These were not primitive ships. When Europeans first arrived in the Straits from the fourteenth century, they were at once impressed with the size and technical advancement of the best of these ships. Together with the smaller *prau* they were to remain the basic cargo ship in the Straits for many hundreds of years and the design remains effective today.

The Southeast Asian junk had a larger crew than European ships of a similar size

Figure 5.1 Characteristic ship types in the Straits, sixteenth–eighteenth centuries (*source*: Kemp, 1980; Reid, 1993).

and, as Reid notes (1993, 49–51), the internal structure and organisation of the typical vessel was often rigidly controlled by a range of regulations. A captain-merchant or *nakhoda* controlled the ship with a master and assistants to oversee maritime matters. The internal cargo space was usually divided into a series of compartments, each of which contained goods to be sold on the voyage by individual merchants. The merchants then, 'usually travelled with the ship, traded his own goods, and bore full risk of shipwreck paying as rent . . . a percentage of the value of the goods with which he set out' (Reid, 1993, 51). Such ships, especially those built by Javanese shipbuilders, could be enormous. Manguin (1980, 267) estimates that around 450 tonnes was not exceptional for Straits junks in the early sixteenth century.

That such vessels, singly or in groups, plied their trade in the Straits and beyond is evident from a range of contemporary chroniclers. Southern China and southern India were the trading destinations of choice from the thirteenth century onwards, but Aden and the Gulf were also noted among their destinations. Coasting, that is, navigating within sight of land, was most common, but there is little doubt that more advanced forms of navigation were practised well before the arrival of what were once thought to be superior western navigational techniques. It is now clear that this myth of European superiority is simply not substantiated. It may well be that the Chinese compass was known and used in the region before the coming of the Europeans (Reid, 1993, 43), and we know that detailed charts of the seas of these 'lands below the winds' existed when the Portuguese first anchored off Malacca in 1511, although the best of these was subsequently lost at sea. Indigenous skills in both short- and long-distance navigation were well developed in the fifteenth century. 'Europeans', notes Ungar (1996, 32), 'were . . . impressed with what Muslim and Chinese sailors could do when they met them in the Indian Ocean'. Whilst the Europeans had the firmer theoretical base in navigation, the pragmatic, empirical skills of Straits sailors were clearly acknowledged. The two elements were quickly synthesised.

It is perhaps tempting to exaggerate the impact of European ships and maritime technology on the societies of the Straits because of the apparent speed with which they seized control of the region. If indigenous ships and methods were so well suited to the trading and military activities that characterised the Straits, why was the European conquest seemingly so fast? It is clear that the organisation and technology of Portuguese naval forces undoubtedly allowed then to make speedy conquests in the region, advancing along a series of stepping stones towards the fabled riches of the East. Aden, Goa and Ceylon were taken in turn before Malacca fell in 1511.

The classic carrack, which was the basis for the European galleon, the building block for the ships used by the explorers and their armies that voyaged east, was based on a skeleton frame and the use of twin masts. The development of a combination of a lateen sail, good for sailing close to the wind, and the traditional square sail allowed a good deal of manoevrability for larger ships. Many galleons were quite sleek, with an average length to beam ratio of about three to one, sleeker than many of their counterparts elsewhere (Kemp, 1980, 34). By the early fifteenth century, European shipbuilders had developed three-masted ships. This evolution, notes Ungar (in Hattendorf, 1996, 43), 'was superior to any European ship which came before it. It formed the basis for European ship design for the next four hundred years and more. It was the basis for the superiority of Europeans at sea.' The caravel, developed by the Portuguese from the late fifteenth century, shared many of the basic features which had placed Iberian shipping at a distinct advantage. It was designed as a long, narrow boat for coastal exploration and messenger services, rather than the wider, hulk-type ship which was the staple of the cargo trade. With a combination of both square and lateen sails it was an excellent, maneouvrable, multi-purpose ship.

The use of explosives was to make a major contribution to the speed of the European impact. Here, the Europeans did hold a decisive technological lead. Explosives were not unknown as a means of making naval warfare more effective amongst indigenous Straits navies, but what marked out the Europeans was that they possessed ships which were, in effect, floating artillery platforms, rather than ships which happened to have a few supplementary cannons. Using these large ships, the Portuguese could anchor off port and shell enemy harbours almost at will. It was this ability to muster key artillery skills which was instrumental in their seizure of key ports along the routeways to the East. Skilled gunners could sink junks and galleys with some ease in harbour. In fighting at sea, the advantages were less decisive; some of the heavier Portuguese ships had difficulty catching native *dhows*, junks and galleys in flight. As the Portuguese control of the Straits was extended through the sixteenth century, they began to adapt their own ships to the demands for flexibility, speed and manoeuvrability in the narrower channels of the seas. Conquest and consolidation each required rather different sets of skills and resources.

Once the European powers had established a form of conquest in the Straits, one of the major impacts on shipping technology was the development of the large trading ships that were to link the fragmented production sites of the East Indies with the ports and markets of Europe. Much of this long-distance trade was dominated by the ships of the great companies, the so-called East Indiamen. The classic ships of this class built on the expertise of European shipbuilding. They were large, often ornate ships, usually equipped with a tall, three-masted rig and, according to Kemp (1980, 40), ranged from an average of around 1,500 tons in the mid-eighteenth century to an average of 2,000 tons in the early nineteenth century. Usually travelling in convoy for security (this was particularly true of the Dutch ships), these fleets brought the products of the East to the European market. They were the foundation on which the fabulous wealth of European merchants was built.

The impact of these large, long-distance trading ships on the commerce of the Straits was immense. Whilst Chinese, Javanese and Malay junks had travelled over huge distances in the pre-European period, it was the scale of the European operations, coupled with an increasing European monopoly over marketing (a monopoly which was essential for the maintenance of high prices to consumers), which made the fundamental difference. It led, for example, to the development of bigger port facilities in the main collecting ports—Malacca for the Portuguese and, in particular, Batavia for the Dutch. At such ports, the concentration of the annual fleet was highly important even where, as in the case of many of the Dutch convoys, they did not sail through the Straits but rather went along the trade winds route to the Cape (Bruijn,

1980). It led to the development of important wharehousing facilities (the so-called *godowns*), where goods could be collected, sorted and packed, as well as a range of administrative and financial houses. Especially important was the way in which local traders and trading networks increasingly fed into these long-distance patterns. The network of indigenous collecters and traders was increasingly absorbed into global trading patterns through Malay and Chinese intermediaries acting at the increasingly international ports of the region such as Malacca, Aceh and Batavia.

CHANGES IN THE INDUSTRIAL ERA

By the late eighteenth century, the impact of the great Dutch and English trading companies was immense. In terms of the size of their fleets and, more importantly, the degree of control they were able to exercise over the centres of collection and production of goods, they had transformed many aspects of the human geography of the Straits. A superior technology allied with organisational skills, heavy capital investment and the judicious application of military force were the keys to the success of European intervention. But those forces remained to some extent superficial. The numbers of Europeans in the region was always small and European influences in the administration, economic development and built environment of the region were confined largely to the towns. If the culture of many of the cities, especially the capitals of Malacca and Batavia, were European, these represented something of a veneer, a superficial imprint on a countryside and region which remained firmly in indigenous hands.

This was to fundamentally alter in the industrial era which followed the hey-day of the great merchant adventurers. With the onset of industrialisation in Europe from the mid-eighteenth century and in the United States in the nineteenth, mercantile gave way to industrial capital and the scale and impact of European intervention in the Straits region changed. Three particular processes can be identified. First, there were major changes in demand for the products of the region, largely fuelled by population growth in Europe, the United States and Japan and by the rising demands for industrial raw materials. Linked with the growth in industrial output in Europe was the need to find secure markets for industrial products outside Europe. The markets of Southeast Asia and China were one such important outlet for the products of the European factories. Imperial might allied to commercial monopolies were a potent force used by the British and Dutch alike.

Second, there were important changes in the mode of organisation of European intervention. The decline of the monopoly Dutch and British East India Companies from the late eighteenth century reflected the collapse of mercantile capital: whilst these monopolies lingered into the nineteenth century, the demands of industrial capital were reflected in new trading arrangements, the rise of joint stock companies and new means of financing investment in the region. The rise of new shipping companies and import–export agency houses, and the development of banking systems were important institutional arrangements which impacted on the geography of the region. They were especially significant in tying together international and local trading and capital circuits through the rise of indigenous middlemen and the so-called 'country traders' of the eighteenth and nineteenth centuries.

Third, there were a range of innovations in the narrower, technical sense which transformed many aspects of life in the region from the mid-nineteenth century. The impact of steam is an obvious first one, associated too with the development of iron ships. Other changes were also significant. From the opening of the Suez Canal in 1869 to the construction of railways in both Peninsular Malaysia and in the tobacco plantations of Sumatra, technological innovations had an important effect on the production, transport and marketing of a range of goods in the region. They helped to initiate the era of industrial imperialism in the Straits region.

Underlying these changes, then, was the growth of demand for tropical goods in the industrial centres of western Europe. The population of western Europe had almost trebled during the nineteenth century, a growth which helped to fuel demand for goods. Timber and tobacco and, later in the century, rubber and tin typified the kind of products sought out by European industrialists. Specialist goods such as spices and precious metals had, of course, long been important products from the region and had fuelled the first phase of European intervention. But these trading monopolies were broken by the mid-eighteenth century and, in any case, by the mid-nineteenth century, supply was no longer a major problem. But the onset of industrialisation created new demands which could be met only through an intensification of production sites in the region. As demand for tea, coffee, timber and metals such as tin and copper rose in the nineteenth century, a rise in both the production of these goods in the Straits region and an intensification in shipping of such goods through the Straits was evident. Three examples will serve to illustrate this.

The tobacco industry in the Deli region of East Sumatra developed from the 1860s onwards under the impetus of Dutch and, later, European investment. By the 1890s, large parts of the Deli region had been transformed, with vast plantations specialising in raising tobacco for the European and American markets. Plantation systems based on careful demarcation of planted areas, a reliance on imported, largely Chinese, labour and the development of an infrastructure of small railways and linking roads dramatically altered the appearance and social structure of large parts of East Sumatra. The capital, Medan, and its port, Belawan, through which the tobacco was shipped, were important nodes of development in the region (Pelzer, 1978). As Table 5.1 shows, by the 1890s the industry had largely peaked and, from the turn of the century, tobacco was increasingly replaced by rubber and, rather later, palm oil. But the impact of tobacco was huge, both in terms of the landscape and the socio-economic structure of the Deli region (Breman, 1989).

The second example, that of the rubber industry, has been well described by Voon (1976). Rubber was increasingly in demand on the European market by the end of the nineteenth century, a demand which was to be massively increased with the development of the internal combustion engine and the motor car. The seeds of *hevea brasiliensis* smuggled out of Brazil by Henry Whickham in 1876 provided the basis

Table 5.1 Tobacco and rubber plantations in East Sumatra, 1864–1932

Tobacco		*Rubber*	
Year	*Number*	*Year*	*Area planted (ha)*
1864	1	1904	651
1874	23	1909	21,926
1884	76	1915	103,112
1886	104	1925	118,875
1888	141	1932	284,213
1891	169		
1894	111		
1904	114		

Source: Breman (1989, 65); Pelzer (1978, 54).

Table 5.2 The growth of the rubber industry in Malaya and Sumatra, 1906–1914

	Number of new companies			
Region	*1906*	*1908*	*1910*	*1913–14*
Selangor/Perak	23	4	40	4
Johore	2	1	11	1
Rest of west coast	10	4	16	3
East Sumatra	1	6	26	3

Source: Voon (1976, 40, 46).

for those rubber seeds sent to Malaya and Singapore in the 1890s. By the turn of the century, planting of rubber had greatly intensified in Malaya and Sumatra, and rubber plantations began to spread along the east coast of Sumatra, often supplanting earlier tobacco plantations and in the states of Malacca, Selangor and Perak in Malaya (Table 5.2).

Dutch, British and American investors planted heavily, instituting a series of rubber booms in the first two decades of the twentieth century. In Sumatra, the residencies of Langkat, Serdang and Asahan were planted heavily by 1914, and the east coast of Sumatra became an important destination for increasing numbers of Javanese migrants. Javanese migrants and Chinese coolie workers also arrived in considerable numbers in Singapore in the two decades either side of 1900 to work on newly opened plantations in the Straits region. As Chapter 7 will show, rubber was an important component in the exports of ports such as Singapore and Malacca. Alongside those burgeoning exports, a range of machinery and finished goods flowed into those ports to cater for the rising demand for the industrial products of the west.

Tin, alongside tobacco and rubber, provides a third example of the impact of western demand on the products of the Straits region. Long known and mined in the region, the tin mines of Perak and Selangor expanded output very rapidly from the

1860s onwards (Tregonning, 1967). As Chapter 4 showed, methods of extraction were often crude, dangerous and environmentally damaging. Nevertheless, large quantities of tin were exported through the smaller ports of the Straits for eventual re-export through Singapore, providing important revenues for Chinese and, from the latter part of the nineteenth century, European investors.

The pattern of primary exports which had long characterised the Straits region was thus further consolidated during the era of industrial capitalism. It was a pattern repeated for a whole range of products—timber, cocoa, tea and coffee, palm oil, hydrocarbons—which has persisted through to the present day. The rise of Singapore as the main exporting and importing point for the region, a rise examined in Chapter 7, was closely connected to the technical changes heralded by the Industrial Revolution in Europe, the United States and, somewhat later, Japan.

Alongside these highly visible sets of changes came a range of less obvious, but no less significant, institutional changes in the trading and business environment of the Straits. The decline of mercantile capital and the monopoly powers of the great trading companies was an important event in the Straits. The monopoly that the English East India Company had on the China trade was broken in the 1830s, whilst the role of the Dutch East India Company was effectively usurped by the Dutch state when that company was virtually bankrupted by the early nineteenth century. In the Dutch sphere of influence, however, the collapse of the company saw its replacement by what was effectively a state monopoly. The creation of the *Nederlandse Handlesmattschappij* in 1824 established a single agency to handle imports and exports in the Dutch sector. In Sumatra, the key products—rubber, tobacco and tea—products which had expanded steadily from the late nineteenth century, were produced and marketed under virtually monopoly conditions. In Medan, from the 1830s, the Culture System of forced cultivation was operated for some four decades to maintain output and exports of these valuable commodities. British, American, as well as Dutch investment poured into these zones (Tate, 1979, 50). Similar investment flows transformed many of the coastal areas of the peninsular into plantation zones.

Alongside the plantations, a range of innovations in the organisation and use of capital was evident from the mid-nineteenth century. The joint stock company, quoted on European and American stock exchanges and drawing on the capital resources of those regions, was the most important vehicle of development, especially as the tobacco and rubber plantation sectors developed. Coupled with these changes in the financing and control of European investments was the increased integration of shipping functions into the wider economic development of the region.

Shipping services through and within the Straits were a mix of private and government initiatives by the end of the nineteenth century. The long-distance trade was dominated by the British and Dutch. A range of British shipping companies were involved in trading through the Straits—many had originated to serve the important China trade (especially in tea) but, as the century wore on, became increasingly involved in the transport of commodities originating in the Straits region. Thus the Peninsular and Oriental (P and O) company owed its early growth from the late 1830s to important mail contracts with the British government to Portugal (hence,

Peninsular), which replaced the old packet services which were often inefficient and corrupt. By the 1840s, P and O had established mail links to Ceylon, Penang, Singapore and Australia. In 1852, it established the first mail steamer service to Australia, passing through the Straits and using the bunkering services of Singapore (Cable, 1937, 118). P and O was required to compete with a growing number of other lines using these waters, especially as the use of steam expanded from the 1860s. The Blue Funnel Line, the Ben Line and the Shire Line expanded their services through and within the Straits. For the Dutch, largely outcompeted by aggressive British lines, the solution was state intervention and, in 1888, the *Koninklijke Pakatvaart Maatschappij* (*KPM*) was set up to try to capture the bulk of Dutch produce shipped from Sumatra and Java (Allen & Donnithorne, 1957, 26). French interests also saw the establishment of a French shipping company, *Messageries Imperiales*, in the mid-nineteenth century, whilst German interests were developed by the creation of *Norddeutscher Lloyd* in 1884, which was especially active on the Singapore–Penang route (Tregonning, 1967, 10–11).

This range of international companies, however, relied on a huge network of local, feeder services to ship products to the main warehouses in Singapore. Many were in Malay or Chinese hands and played a crucial role in feeding produce into the main companies. One of the most important local lines, the Straits Steamship Company, was set up in January 1890, drawing predominantly on Chinese capital. As Tregonning (1967) notes, it was to become one of the most important lines in the region, with a major network of feeder routes for moving both goods and people around the Straits and feeding in to the international routes.

THE IMPACT OF THE NEW TECHNOLOGY

The nineteenth century saw major innovations in both shipping and in the routeways of the Straits region, innovations which, in combination with increased demand for products and new modes of organisation, helped to transform the geography of the region. By the early part of the century, shipping advances had produced a range of fast, sleek sailing ships which specialised in the speedy transport of goods, especially tea, from China, through the Straits and onto the European market. These were the 'clipper' ships, which were designed above all for speed. British clippers of the 1830s and 1840s had a length to beam ratio of between seven and eight, were often built around a metal frame which allowed for much larger (and hence more cost effective) ships to be built. The most famous of these clippers could sail from China to London in around 100 days. Their zenith, around the middle of the century, was to coincide with two major changes.

In 1869, the opening of the Suez Canal greatly shortened the sea-routes to the East by largely eliminating the route around the Cape of Good Hope. It also eliminated much of the costly infrastructure surrounding the overland route across Egypt which was heavily used by companies such as P and O. That company, for example, had been required to both keep up the overland routes and hotels through Egypt, and

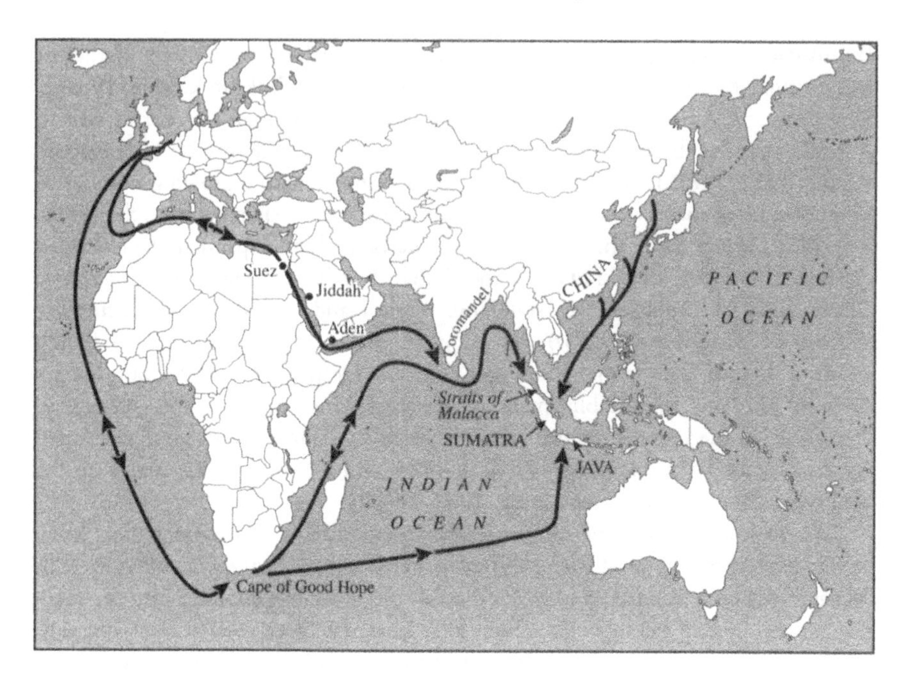

Figure 5.2 Pre- and post-Suez trading routes to the Far East.

maintain both 'Eastern' and 'Western' ships at either end of the route. The opening of the canal eliminated the need for these costly investments and meant, furthermore, that one ship could make the entire journey out East and back. The opening of the canal also increased shipping using the Straits of Malacca rather than the Sunda Strait, which had previously proved attractive to ships taking the trade winds from the Cape of Good Hope towards the coast of Australia, before sailing north towards Java and southern China. With the opening of the canal, the new steamships could journey from China to London much more rapidly than the best clippers using the Cape route. The clippers continued to be used through to the early twentieth century, largely in the Australian wheat and wool trade, but the advent of the Suez Canal and steam sounded their death knell.

The development of the steamship was critical in the commercial life of the Straits, but its arrival was slow, piecemeal and heavily dependent on innovation. The earliest steamships from the 1820s were heavy, cumbersome and uneconomic. But the development of the single cylinder engine in the 1840s and the compound engine in the late 1850s allowed for greater power per unit of coal. They increasingly helped to overcome the fundamental problem of freight transport by steam: how to balance the need to store large quantities of coal with the requirements for storage of freight. Early steamships were made of wood. By the 1860s, iron steamships had been developed which, coupled with the opening of the Suez Canal, began to open up the possibilities of long-distance sea freight under steam. The development of steel ships from the 1870s gave further impetus to steam at the expense of sail. In 1868, a census of global shipping showed 4.6 million tons of sailing ships against only 0.8 million tons of steamships. By 1872, sail tonnage had dropped by 0.5 million, whilst steam tonnage had more than doubled (Kemp, 1980, 51). As the efficiency of marine engines grew, and as steel replaced iron as the material of choice for steamships, more and more owners and companies made the switch. But, especially for the local and regional lines, the switch was slow and piecemeal and the shipping returns for Singapore, Penang and Malacca continued to show the importance of sail ships well into the twentieth century. The harbours of the Straits continued to be crowded with a jumble of sailing craft well into the 1950s and 1960s. Even today they are a common sight on these waters.

The development of steamships required a series of changes in the ports and facilities in Straits harbours. Thus, both individual companies and the port authorities at ports such as Singapore, Malacca and Aceh developed new coal bunkering facilities to encourage ships to bunker there. At Singapore, P and O had its own bunkering facilities for its ships, with a considerable storage area and labour force. In 1905, the newly created Tanjong Pagar Dock Company set about establishing a range of bunkering and repair facilities for the increasing number of steamships calling at the port.

The development of port facilities along the Straits was an important outcome of both the increased commercial penetration of the Straits region through plantation production, and the greater intensity of shipping services in the region. Along the Malay coast a series of small ports were developed towards the end of the nineteenth century to service the growing trade in primary products. Whilst older ports such as Penang and Malacca continued to trade heavily, especially with their hinterlands, new ports were also set up. Thus Port Weld was developed at the terminus of the railway to cater for important tin exports despite an unpromising site. A new wharf was constructed by the Straits Steamship Company in 1928 to cater for its growing trade. Teluk Anson was established in 1883, some 50 km inland up the Perak river at the terminus of the railway catering to the demands of the tin trade. Port Klang was developed as a new port for Kuala Lumpur from 1901 so that the federal capital would not be dependent on Singapore or Penang for its supplies (Tregonning, 1967).

On the Sumatran side, the growth of port and shipping facilities depended, as in Malaya, on the extent of commercial penetration into the hinterland. The port of Banda Aceh remained important through the century, whilst Belawan played an important part in exporting the rubber, tea and tobacco of the Deli region. Further south, the port of Palembang traded extensively, especially with the discovery and development of important hydrocarbon resources in the Jambi-Palambang region at the turn of the century. The opening of the Rantau field in the 1930s was also an important stimulus to port development in the region.

In recent decades the link between technology, resource use and development has remained important. The expansion of hydrocarbon production in Sumatra has been an important stimulus to development both of port facilities and of infrastructure generally. These and related developments in port facilities such as containerisation and storage facilities are examined in Chapter 8.

It is evident that a key factor in the evolution of shipping services and ports in the region was the opening up of the interior of both Malaya and Sumatra to commercial exploitation through plantations or the exploitation of mineral resources. Accompanying this exploitation were important infrastructural developments which helped consolidate the economic power of the major ports, in particular Singapore. The building of roads and railways to link port and hinterland were especially important. In the Medan region, the development of major plantations had been accompanied by the building of a network of small lines linking the plantations with the capital Medan and the port of Belawan (Pelzer, 1978). In Malaya the construction of the railway from Penang, through the tin mining zone of Perak to the capital Kuala Lumpur and on to Johor was largely completed by 1910. Branch lines connecting to the ports of Klang/Swettenham, Port Weld and Port Dickson facilitated the growth of those ports (Tate, 1979, 193). The construction of the causeway linking Singapore and Johore Bahru in 1923 was important in allowing Singapore to capture more of the import–export market of Malaya.

To some extent, then, it is possible to trace the impact of a series of innovations on the landscapes and peoples of the Straits and to identify a relationship between technological change and socio-economic development. The effects of new shipping technologies—from the use of the lateen sail and the cannon, to the arrival of steam and the iron ship—were fundamental in reshaping how the peoples of the region evaluated and exploited their resources. Equally, new methods of organisation—from the East India Companies to the joint stock company to the shipping conglomerate—have been important in the development of the Straits. But, as Chapters 6 and 7 will show, the relationship between technology and development has never been a deterministic one; human ingenuities and abilities have been equally significant in the Straits region.

Part 3

COLLECTIVE HISTORIES

6

EARLY MARITIME EMPIRES, *c.* 700–1700 AD

Forest and sea were the interconnected environments upon which the economic, political and social development of the Straits region were anchored in the past. The forest, with its diverse range of products, ranging from the everyday to the exotic, from sago and bamboos to camphor and precious woods, provided goods for both subsistence and trade. The seas and estuaries of the region were arteries along which flowed both the forest products and the products of the sea—pearls, fish, trepang. Those sea-lanes also brought with them goods from outside the region as well as the peoples, ideas, and religions which were to animate the urban and commercial life of the islands of Java and Sumatra, the many islands of the Straits region and the Malay peninsula. The interactions of those environments, coupled with the geo-strategic importance of the Straits themselves, created a range of opportunities for the development of economic and political life.

Probably the defining feature of the diverse states, cities and kingdoms that emerged in the region was an openness to a range of external political, economic and military influences. By contrast with the great, landed kingdoms on interior Java or the Khmer in Cambodia, built on the agricultural products and rhythms of its territories (Osborne, 1988), those of the Straits were usually more open, fluid and cosmopolitan. Influences from India, China, the Arab kingdoms and, somewhat later, Europe, were important in shaping the character of social and economic life. Ultimately, such openness, a product of a location abutting the major trade routes, meant exposure to armed struggle, conquest and political and economic insecurity. The port cities that were to play such an influential role in shaping the historical geography of the region—cities such as Malacca, Pasai, Aceh and Johor—were, by their very structure and location, especially vulnerable to attack and conquest.

EARLY PEOPLES

We know relatively little about the indigenous peoples of the region—the early sailors, traders and harvesters who first settled along the shores of the Straits. Partly this is because they have left few traces—either archaeological or literary—but also because the varied ethnic groups of the region mixed together speedily. It is also the case that much of our knowledge of these groups is based on often sketchy European ethnographic and anthropological accounts, accounts which often emphasised only those features which were most striking to the colonial mind. As Sopher (1987, 296–306) has emphasised in relation to the sea nomads of the region, such accounts have to be used with great caution.

There is evidence to suggest that well before the rise of Srivijaya in the eighth century, an extensive trading network already linked the *orang asli*, the interior groups of Sumatra and Malaya, with the sea people or *orang laut* (Andaya & Andaya, 1988, 10–13). This interaction, built on the exchange of commodities, also fed into a developing coastal trade. From earliest times, noted Wheatley (1961, xxi), 'gold was the lure which drew Indian traders eastwards' and the gold deposits which extended from Patani, through western Pahang and eastern Negri Sembilan and down to Malacca, were much sought after by Arab, Indian and Chinese trader alike. Supplying that demand drew a range of indigenous peoples into contact with a nascent maritime economy in the Straits. Alongside gold, tin was an important commodity from earliest times. Again, we know from some of the earliest Chinese sources that tin was sought after and traded in a range of settlements in the region. Wheatley (1961) has emphasised how its importance was reflected in the writings of many Chinese and Arab chroniclers. According to Andaya and Andaya (1988, 12), 'from the fifth century AD it may have been shipped to India to be used in alloys like bronze for the manufacture of religious images'.

Coupled with these products, a range of other goods in the region played a role in both local and regional trade, although it is difficult to be precise about the size and geographical extent of that trade. But there is enough evidence to show that outside traders—Chinese, Indian and Arab—were both aware of and skilled in securing supplies of important jungle and sea products such as precious woods, coral and pearls. In exchange, these traders brought textiles (particularly from south India), pottery, glassware and beads. Along the shores and hinterland of the Straits, we can be confident that networks of trade had developed at an early stage.

Given the existence of this trade, what of the inhabitants of the Straits region? Sopher (1987) has argued that, in pre-Moslem times (Islam arrived around 1400) the peoples of the region were assiduous in taking advantage of the productive coastal environment for trade and enrichment. As Chapter 4 suggested, the sea-strand region was rich in potential trade goods on both sides of the Straits, goods which were seized on by the varied *orang laut*, who were able to use their skills as divers, fishers and general scavengers of the coastal region to develop often intricate trading networks which later groups were able to tap into. The ecology of the Straits—warm waters, low salinity, a relatively low swell by comparison with the western coast of Sumatra—provided good conditions for harvesting these marine resources. Sailing was fairly safe and straightforward.

The numerical importance of both *orang asli* and *orang laut* must have been considerable in pre-Moslem times; it was a population much stronger than the apparent 'rump' population that survived into the twentieth century to fascinate colonial observers. For Sopher (1987, 7), these peoples, once widespread in the region, probably came from one cultural hearth: 'although all the sea nomad groups have experienced cultural modification, it is argued that they belong to one original cultural stock'. The arguments for and against this diffusionist model of cultural groups are beyond this present chapter, but what is important to emphasise is the wide geographical spread of these groups, and their cultural impact throughout the region. It is also important to stress their role as traders, a role which prefigured some of the later developments in the Straits.

The *Celates*, for example, a group dominant on the Malay coast of the Straits, has been seen as playing an important role in the founding and consolidation of Malacca; whilst other groups, such as the *Bajau Laut* from the Riau archipelago, were important in later economic and political developments. The role played by the foot-loose *Bugis*, originally from the seas around Sulawesi but highly migratory from the seventeenth century onwards, is perhaps a later example of the kind of role the sea nomad groups played (Pelras, 1996). For Sopher (1987, 344), both Chinese and Arab sources point to a range of activities by such groups, some peaceful, others piratical, back to at least the first millennium.

We would argue, then, that long before either the rise of Malacca or the arrival of European traders, the Straits region was home to a diverse and, in places, economically important network of trading relationships linking indigenous interior groups with their counterparts on the coast who, in turn, linked with foreign traders. These networks criss-crossed the Straits from one side to another in search of particular products, hopping equally from island to island, especially at the southern entrance to the Straits. It was from this basis that the important trade goods began to enter into an international trade from the sixth or seventh centuries onwards.

Indian traders were an especially important element in the early trading contacts of the region's peoples with the outside world. As SarDesai (1994, 16) has argued, 'the overwhelming Indian participation in East–West trade brought large numbers of Indian seafarers and merchants to Southeast Asia, where the rulers were also the principal traders'. Unlike the better documented China trade, the precise nature of the products traded is unclear. To the Indians of the first century, Southeast Asia was commonly known as Suvarnadvipa or Land of Gold suggesting an early and obvious interest in precious metals, found and traded in the Straits region, particularly Sumatra. Making their first footholds north of Kedah, traders from Gujerati and the Coromandel coast brought textiles to exchange for local supplies of metals and forest products.

As goods flowed, so too did people and ideas. A process of acculturation, of the steady absorption of Indian ideas began to take place. The people of the Straits were put in contact with Hinduism and Buddhism, as well as with a range of introduced notions of political and state organisation (Sandhu & Wheatley, 1983, 8–13). It is interesting to note the slow, syncretic way in which Indian ideas—religious ideas as well as others—began to permeate the region over the middle centuries of the first millennium. It was a slow, almost evolutionary process quite unlike the sudden, dramatic changes which accompanied the arrival of Islam from the early fifteenth century. But, like Islam, this diffusion of ideas and ideologies deeply marked the societies of the region and reflected its openness to the world at large over the centuries.

Alongside India, China was the other key influence on the region. Its influence was not steady for the openness of China itself waxed and waned. Trade with the outside world depended overmuch on internal political conditions. But the importance of the China trade for the societies and cultures of the Straits was considerable. Certainly that trade is much better documented than is the case for Indian merchants, primarily because Chinese governments sought to control and channel the products and wealth of that trade to their own ends. By and large the countries of the great Southern Ocean, the Nanyang, were required to trade through a system of tribute in which trade goods were sent through official missions to the Chinese ports. Political relations, especially a willingness by states in the Nanyang to acknowledge the overlordship of China, proved to be the key to unlocking those trading connections.

The early development of that trade is traced in Paul Wheatley's classic, *The Golden Khersonese*, first published in 1961, whose very title alludes to a fabled land of riches and wealth, the scene of Chinese traders from at least as far back as the fifth century AD. 'The increasing use of the sea to transport goods between western Asia and China', wrote Wheatley (1961, 17), 'created an environment well suited to the rise of ports in numerous places in the Malay archipelago.' The increase in the volume of trade, both from India and China, not only stimulated the economic activity of the indigenous peoples of the Straits, it also led to the establishment of ports and trading places. The character of the monsoon winds further emphasised the locational advantages which were an important factor in the rise of the Straits ports.

EARLY EMPIRES

The establishment and structure of the most significant early states and empires in the Straits were closely linked with the role the region played as a cultural, political and economic crossroads. The trading networks which linked Sumatra, the archipelago and the islands of the Straits with the distant centres of India and China were crucial in establishing the conditions which facilitated the emergence of powerful polities towards the end of the first millennium. The kingdom of Srivijaya was the most important of these. Srivijaya first appears as a political and economic force towards the end of the seventh century and it played a dominant role in the Straits until well into the thirteenth century. Its precise origins and structure are hazy and conjectural—since the 'discovery' of Srivijaya by the French historian George Coedes in 1918 (see Coedes, 1966; Wolters, 1967), there has been much debate about the kingdom.

Its geographical base was in southern Sumatra, centred almost certainly around the present-day city of Palembang, some distance inland along the Palembang river. Its location on the path of the northern monsoon trade routes gave it a pivotal place in the China trade and, from its inception, her rulers were assiduous in cultivating good links with China through the prompt payment of tribute. Since for much of this period, the sending of tribute ships was the only legitimate way to trade with China, the acknowledgement of 'vassal' status was important to economic success. Between 960 and 983, for example, Srivijaya sent no fewer than eight missions to China to cement its good trading links with the major trader in the region (Andaya & Andaya, 1988).

The rise of Srivijaya was built on firm control of the trading groups of the region. Given the nature of the terrain and the limited military resources of the state, territorial control was always of only secondary importance. The control of key ports, good relations with the peoples on both sides of the Straits and a degree of naval presence in the Straits were the prerequisites for success. The *orang laut* were especially important: they were the key to unlocking the collection and sorting of many of the valuable sea, strand and forest products of the region which were traded on the China market. In addition, if these groups had a clear economic interest in trading through Srivijaya, they would also help in ensuring a degree of free access to the sea lanes of the Straits. Their piratical role could, with sensible management, be turned to the advantage of the state itself. The cooperation of these groups was crucial in ensuring the success of Srivijaya.

Geographical location alone was hardly sufficient to explain the triumph of the empire for its key capital, Palembang, was some 120 km from the Straits themselves. As Wheatley notes (1961, 293), 'only after she had established herself as a south-east Asian power could she reap the full fruits of her nodality in the Nan-Hai, and this she had achieved by the last quarter of the seventh century when her territory extended over the southern half of Sumatra, the island of Bangka and possibly part of western Java'.

Alongside this territorial core, Srivijaya also secured significant vassal–client relationships with a range of smaller kingdoms in the region. At least half a dozen smaller states from Patani and Kedah to Johor had, at some stage, formed part of these wider systems of economic and political exchange linking the periphery to the core at Srivijaya. By the beginning of the eleventh century one Chinese source noted that at least 14 cities in the region paid tribute in one form or another to Srivijaya. These cities were on both sides of the Straits emphasising the extent to which the Straits functioned as a net, drawing ports and kingdoms together, rather than as a barrier, setting off developments on one side from those of the other.

The rise and apogee of Srivijaya, and its ultimate decline by the end of the thirteenth century, reveals much about the economic and political conditions which animated the region as a whole. The importance of long-distance trade remained axiomatic. As in western Europe at this time, it was long-distance trade which laid the foundations for the accumulation of surplus value in the institutions (political, economic and cultural) as well as the built form of the city. The China trade, and to a lesser extent the India trade, was the base on which Srivijaya's wealth was anchored. Alongside this trade there followed a range of cultural, religious and political currents which helped to define the character of Srivijaya, its laws, culture, religion and built form. Beneath that long-distance, 'exotic', trade was an important base of local trade, built on a network of vassal states linked by a network of local, indigenous sailors, shippers and traders which fed into the wider connections of regional interchange.

The nature of the Srivijaya polity also underlines the significance of personal alliances in the exercise of commercial hegemony in the Straits. By contrast with European conceptions of power and the state, control of territory was much less important than control of people and routes. Power and authority rested not so much in military force and military forts but rather in networks of personal links and relationships, built on mutual self-interest in the benefits of stable trade. The power of Srivijaya, then, like that of many of the powers that followed it—Malacca, for example, or Johor—rested on its ability to forge alliances with local and regional powers, alliances which allowed it to control the nature—and profits—of trade. The collection and sorting of local forest and sea products were thus tied, through mutually beneficial systems of alliances and tribute, to a wider power with the ability to sustain trade on a higher geographical scale. Once these ties were weakened by changed economic or military conditions, the authority of the centre was fatally weakened.

The end of Srivijaya's power reflected the collapse of many of the features that had contributed to its rise. The ability to control and police local and regional trade came under threat from the increased opening up of trade with China from the end of the twelfth century. During the late Sung and Mongol periods, the tight restrictions on trade with China, regulated largely through the tribute system (although enterprising merchants could take the risk of by-passing these restrictions), lapsed and, as private trade with China grew, the raison d'être of Srivijaya was threatened. Smaller port cities could now by-pass the larger city and trade on their own account. By the thirteenth century, a network of kingdoms in both Sumatra and the archipelago were trading directly with cities in China—Barus and Kampe in the former, Kedah, Perak, Pasai, Tengganu and Pahang in the latter. It also appears that the central authority of Srivijaya declined due to both internal dynastic disputes and military challenges in the eleventh century. Coupled with the rise of the inland Javanese empire of Majapahit, an empire with a strong interest in the Straits trade, the power of Srivijaya declined. As Reid notes (1993, 203), by 1300, 'the Straits of Malacca area, long the stronghold of the SriVijaya empire, was contested between Javanese, Chinese, Tai and local Sumatran forces'.

THE RISE OF MALACCA

The coming to prominence of the port city that was to give its name to the Straits in the early years of the fifteenth century heralded a period of between 100 and 150 years when a Malay city was to dominate the economic and political life of the region. By 1511, when the city fell to the Portuguese, its particular pattern of economy, government and cultural life had become central to later Malay conceptions of state and kingdom. Historically and culturally the importance of the city, both at the time and in later evocations of the Malay culture cannot be underestimated.

Much debate has centred around the creation and early years of the city. Two major and sometimes contradictory sources form the basis of most analyses. The *Sejarah Mélayu* or *Malay Annals*, was probably first written down in the seventeenth century as a part true, part mythic, account of the rise of the city. Alongside this is a major European source, the *Suma Oriental* (*Complete Treatise of the Orient*) written by Tome Pires, a Portuguese apothecary sent to Malacca in 1512, which gives an account of the history, trade and life of the city. There are important debates about these sources and their value in the historiography of the region which are examined in, for example, Andaya and Andaya (1988, 37–39) and Wheatley (1961, 306–309). What these accounts suggest is that by around 1400, *orang laut*, perhaps together with groups of the *Celates* peoples (Sopher, 1987, 319–320), had established a small port around the mouth of the Melaka river. The position was strategically important. Whilst not at the narrowest point of the Straits, it was in a position to control a narrow navigable channel through which much of the trade of the Straits would pass. Although not possessed of an especially favourable sheltered port, the roadsteads outside the city were reasonably secure, and the river itself was navigable for some distance inland which allowed for the development of basic berthing facilities. The hill overlooking the port provided a good site for a defensive position.

The slow decline of Srivijaya and the rise of Malacca were not coincidental. Rather the growth of Malacca represented a degree of continuity in the characteristic port city of the Straits, for many of the factors which had benefited the rise of Srivijaya operated in the case of Malacca. According to Reid (1993, 14), the early fifteenth century saw a significant rise in trade in the region generally. There was, for example, a minor boom in demand in China for a range of Southeast Asian products around this time. As Wang Gungwu (1964) has shown, in the first decade of the fifteenth century, a number of Chinese sources emphasised how individual officials and expeditions of the Emperor Yung-Lo (1402–1424) visited Malacca in search of tribute and trading opportunities. As early as 1405, it is argued, 'Muslim traders had informed the [Chinese] court that Malacca was a flourishing commercial centre' (Wang Gungwu, 1964, 100). This new, favoured relationship with China was assiduously cultivated by Parameswara, the ruler of the city, and was to be a key element in the commercial success of Malacca. In the course of the fifteenth century, a total of 29 tribute missions were undertaken from Malacca to China. Only Java, Siam and Champa sent more missions in that period (Reid, 1993, 16).

A key feature of the emergence of Malay Malacca was the fact that, like Srivijaya before it, its sphere of influence united rather than divided the Straits. In channeling its growth, the city drew on both the products and political loyalties of a range of states on both sides of the water. 'It is evident,' note Sandhu and Wheatley (1983, 509), 'that the kingdom as a whole constituted a galactic, patrimonial-style state,' with its core in Malacca and its periphery scattered over a wide geographical range. As with Srivijaya, military-style control of territory was largely irrelevant. What mattered was the structure of vassal–client relations, structured as much by mutual

Table 6.1 Vassal and client states of Malacca in the early fifteenth century

Malay	*Sumatran*
Beruas	Rekan
Perak	Rupat
Selangor	Siak
Kelang	Simpang-Kann
Sungei-Ujong	Inderajiri
Muar	Tungkal
Johor	Lingga
Singapur	
Bentan	

Source: Sandhu and Wheatley (Eds) (1983).

economic self-interest as by the force of naval power. Cooperation with the important groups of the islands and estuaries of the straits was vital to the success of the city. As Table 6.1 shows, a range of states on both sides of the Straits owed varying forms of allegiance to the city. This range of associations with states on both shores undoubtedly facilitated the important entrepot and transhipment functions which were at the heart of the economic success of the city.

By the end of the fifteenth century, the *Suma Oriental* records something of the range of products shipped to and from Malacca. Table 6.2 classes these by product and by destination and serves to emphasise the enormous geographical and product range of the city's trading networks. At one level, the long-distance trading functions of Malacca tied it in with the important Chinese and Indian markets. Flows of spices, precious metals, porcelain and pottery, textiles, jungle and sea products animated this trade. Indian, Chinese and Arab traders found a ready welcome in the city where lines of authority and the costs of trading were clear and acknowledged. At the sub-regional level, short-distance trade was vital. It provided many of the products that entered into the international trade or that were used in the exchange of Indian and Chinese imports into the city. Thus merchants from Gujerati or the Coromandel coast, sailing to Malacca with Indian textiles or perfumes and drugs from Arabia, would exchange these goods at the city for the products of the China market. Alternatively, they might exchange textiles for important Straits products which, in turn, would be further exchanged onto the China market or shipped back to find a market in south India. Trading networks at both the international and sub-regional level were thus complex and interlinked.

Perhaps the defining feature of the city's economy was its ability to provide a safe and secure environment within which these complex transactions might take place. Tolls and taxes ensured that its rulers shared in the prosperity this rising trade brought and, through a system of alliances and vassal–client relations, that those further down the hierarchy shared in this growth. Malacca's most valuable territories were her seas, her most important asset was her ability to channel regional and international trade to her own ends.

Table 6.2 Products traded to and from Malacca, early sixteenth century

To Malacca	*Product*
Long distance from:	
Bengal	Textiles, metals, pottery
Gujerat	Textiles
Middle East	Textiles, drugs, metals, jewels
China	Textiles, aromatics, jewels, ceramics
Regional from:	
Sumatra	Spices, aromatics, slaves, sea products, metals
Java	Metals, spices
Borneo	Metals, spices, sea/forest products
Siam	Spices, jewels, gems
T'ai	Rice, metals
Moluccas	Spices
From Malacca to:	*Product*
Bengal	Pottery, metals, spices, jewels
Gujerat/Middle East	Ceramics, metals, spices, textiles
China	Jungle/forest products, textiles, spices, metals
Brunei	Metals, jewels, textiles
Sumatra/Java	Textiles
Moluccas	Textiles, metals

The nature of authority in the city was thus vitally important to her continued economic success. Clear systems of authority and control were established to ensure that trade could flourish and that traders could be confident that their goods would not be seized during their passage in the city waiting for the arrival of the Chinese or Indian traders or waiting for the monsoon winds to turn. A collapse in central authority—as for example in the last years of Srivijaya—could seriously threaten trading confidence and profits. At the top of the government hierarchy, just below the sultan, was the chief minister (*bendahara*). Below him, the *temenggong* had a key position controlling such matters as law and order and weights and measures in the city. His role was important in ensuring the stable, well-ordered conditions propitious to good trade. Below him were four key officials—the *shahbandars*. Each was responsible for good order in the trade with particular areas—China, Java, Bengal and Gujerat.

What did the city look like at this time? It is difficult to be sure since the main sources are sparse on this subject. Certainly by about 1415 the central authority was located on the banks of the Malacca river abutting the slopes of present-day St Paul's Hill. Along the lower slopes of the hill, the core Malay community congregated, possibly on the pile-dwellings which were characteristic of many Malay dwelling areas at this time. The swampy, muddy riverbank would in any case have supported little else of substance in the way of houses. By the late fifteenth century, it appears that this district may have become a more distinctive commercial centre, perhaps suggesting a degree of residential and occupational segregation. For Sandhu and Wheatley (1983, 532–535), a general model of a core area, with the palace or *Istana*, mosque and government buildings, a retail bazaar and a wholesaling centre, seems to have developed by the early sixteenth century. There do appear to be elements of a segmented land use pattern such as developed in other cities in the region at this time.

The influence of the city on the Straits region and beyond was not only political and economic—it also played a key role in the dissemination of Islam. Moslem merchants, from the Middle East in the early phases, and then from southern India, were important traders in the city and brought Islam to Malacca in the early fifteenth century. The *Malay Annals* note the significance of the arrival of Islam in the life of the city. Across the Straits, the smaller port city of Pasai, at the northern end of Sumatra, had adopted Islam in the fourteenth century and had flourished with the influx of Moslem merchants, money and ideas. The lesson for Malacca was clear and the new city swiftly adopted the religion.

As its empire developed, so too did Islam. It spread to both sides of the Straits, facilitating the development of commerce, culture and language. 'Throughout the fifteenth century,' note Andaya and Andaya (1988, 54), 'Melaka's reputation as a commercial and religious centre established it as the yardstick by which other Muslim kingdoms in the archipelago were measured.' The power and wealth of the city, so vividly described for a European audience in the *Suma Oriental*, contrasted with the paucity of development in the coastal hinterlands. It was a city built on mercantile trade, rather than its own production and produce, and it relied heavily on imported rice and other foods. The hinterland was in any case not especially propitious to agricultural development and it was not until the late eighteenth century that the city began, almost by default, to develop pepper and gambier planting on its own account.

Towards the end of the fifteenth century, the monopoly exercised by Arab traders on the lucrative spice trade from the Far East to the growing markets of Europe was threatened by the increasingly powerful state and merchants of Portugal. Portuguese merchants, having seen the profits that the Venetian Empire had accumulated from the spice trade, began to extend their involvement in the trade. Spain, newly unified under Ferdinand and Isabella, was also developing a range of interests in Middle and Latin America, interests which the Portuguese hoped to replicate in the Far East. The Portuguese were also driven by a strong missionary zeal to proselytise the Christian faith in the lands of the infidel Moslem. It was a heady mix of ideals and greed. As SarDesai notes (1994, 60), 'trade and religion were two sides of the same coin; to deprive the Arabs of trade profits and to kill them as enemies of Christianity became the passion of the Portuguese for the next several decades'.

By the start of the sixteenth century, highly skilled Portuguese navigators had rounded the Cape of Good Hope and were actively exploring the vast expanses of the Indian Ocean. As we noted in Chapter 5, their military equipment, notably the ability to use explosives, had helped them put a stranglehold on the spice trade by seizing a series of forts and strongpoints which could be used to channel the trading routes to the benefit of the monopolistic Portuguese. Forts at Cape Town, Hormuz, on Sri Lanka and on Goa were established. A base in the Straits was an obvious next step.

In June 1511, Alfonso de Albuquerque launched an attack on Malacca and, after two attempts were initially repulsed, took the city, putting the Moslem population to the sword and forcing Sultan Mahmud Syah and his son Ahmad to flee to the island of Bentan in the Riau archipelago, at the southern entrance to the Straits. The Portuguese were confident that their seizure of the city would allow them to speedily monopolise the lucrative trade of the region, forcing ships to call at the port and to pay the appropriate taxes and fees to the authorities. For the Portuguese, the seizure of Malacca was seen as the final thread in the knot that would strangle the power of Venice and secure Portuguese control of the spice trade. But the Portuguese were not aware of the true nature of Malacca's power, which lay less in its specific geographical position and much more in the network of personal and vassal relationships that had tied it to indigenous and foreign traders on the one hand, and compliant states on the other. Without that network, the city was bereft of much of its enduring influence and economic strength. As Lewis (1995, 8) rightly notes, 'the essence of Malacca's power was not its physical location, but its social superstructure, the court and its ruler . . . though Malacca was a useful and favoured site, it was only one of many; the trade did not flow to it automatically'.

In the early phase of Portuguese control the city continued to prosper. Whilst the influential Moslem traders of the region fled the newly-Christianised city, the Portuguese were able to seize a goodly proportion of a growing spice trade with Europe. The Portuguese navy was also reasonably effective in compelling ships using the Straits to call at Malacca. The city was rebuilt and a new fort constructed on St John's Hill to emphasise the power of the Portuguese rulers. By the mid-sixteenth century, one estimate suggests that Portuguese profits from the spice trade, much of which passed through the city, amounted to four times the internal revenue of Portugal itself (SarDesai, 1994, 61). But, from the 1560s, the captain-governors of the city became greedier. They sought to increase port duties and taxes, discouraging merchants from calling there. In addition, the Portuguese navy was finding it increasingly difficult to defend their forts and police their waters. The costs of maintaining a large naval force in both the Indian Ocean and the Straits (the monsoon winds prevented one force patrolling both) were becoming too high. Profits and spice shipments from Malacca peaked in the 1570s and 1580s (Reid, 1993, 21): by the close of the sixteenth century, the power of the city and the Portuguese grip on the trade of the Straits were waning and the profits of the city fell.

Despite the incursion for the first time of European traders and adventurers into the region, it is perhaps evolution rather than revolution which best characterises the trading systems and cities of the Straits. The fall of Malacca weakened only the specific location of that state—many of the traders, merchants and sailors who had frequented the city simply upped sticks and moved elsewhere. Alternative ports, more welcoming to Muslim traders, developed, anxious to control and channel the flow of goods through the Straits. Kedah, Perak, Pahang, Johor and Aceh exemplify the continuities of social and economic life. Aceh was the most significant of these.

THE RISE OF ACEH

The fall of Malacca to the Portuguese was to bring unexpected benefits to the sultanate of Aceh on the northern tip of Sumatra. Many Muslim merchants, unwilling to deal with a Christian power, fled across the Straits and came to Banda Aceh, the capital of an increasingly prosperous hinterland. The city, along with its rival Johor at the southern end of the Straits, was to become a powerful and influential state in the Straits. Its power lasted longer than Malacca—at least through to the mid-eighteenth century—and, like Malacca, its power was rooted in an ability to control, channel and manipulate the trade of the Straits to its own ends. Its sphere of influence extended well into its hinterland—drawing on the wealth of Sumatra—and along and across the shores of the Straits (Figure 6.1).

The products traded through Aceh were familiar ones. Spices from the Moluccas, cloth from Coromandel and Gujerat, precious metals (tin, gold) from Perak and inland Sumatra and pepper. Unlike Malacca, Aceh was also able to benefit from the products of its hinterland. A domestic pepper industry was developed by the Achenese and, as early as 1500, the area around Banda Aceh and neighbouring Pasai had become significant pepper-growing areas (Reid, 1993, 250). The boom in pepper exports to Europe from the 1540s onwards undoubtedly stimulated the growth of the sultanate. As one of the first centres of Islam in the region, the city was able to attract many Muslim merchants and traders into its port. These merchants not only skilfully manipulated the spice trade, they were also important in developing pepper exports from Banien on Java and Lampung in southern Sumatra.

By the late sixteenth century, Aceh was by far the most important pepper market in the region, and its power and influence soon brought it into conflict with both Portuguese Malacca and Johor. Thus Aceh launched a series of attacks on Malacca in 1558, 1570 and 1575, whilst Johor attacked Malacca on at least three occasions in the first two decades of the sixteenth century in attempts to dislodge the Portuguese from their foothold in the Straits. Not surprisingly emnity grew between Johor and Aceh, especially as the latter sought to extend its influence along the Sumatran coast. At its peak, at the start of the seventeenth century, the Straits region was effectively divided between the three major states—Portuguese Malacca, territorially limited but with a strong naval presence, Johor and Aceh, the latter with vassal states extending almost to the southern tip of Sumatra.

Under its most famous ruler, Sultan Iskandar (1607–1636), its economy, court system and political power were considerable. Banda Aceh had a powerful and diverse merchant community and its court pursued negotiations with a number of powers in the region. At the peak of her power, the rise of Aceh illustrates the extent to which political, economic and cultural influences stretched across the Straits rather than ending on her shores. Thus Aceh was heavily reliant on the tin trade of Perak, it drew,

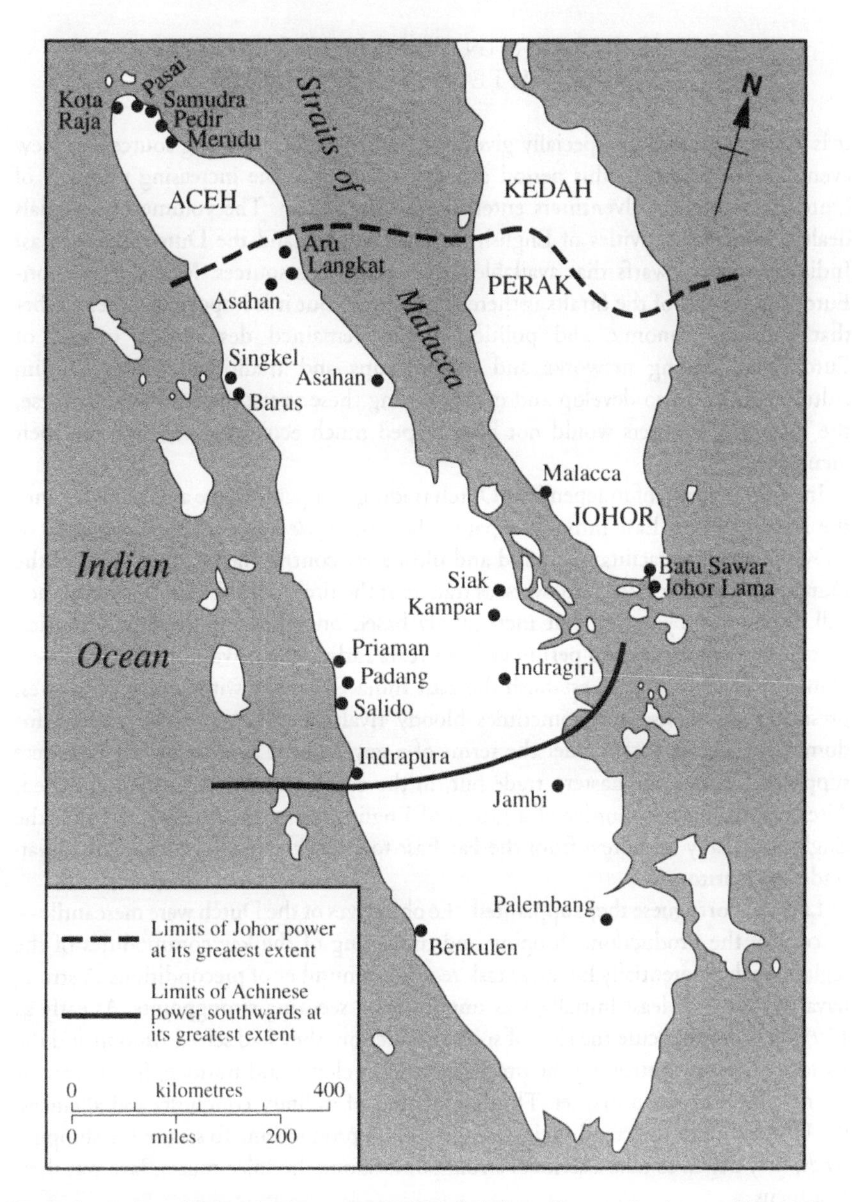

Figure 6.1 Aceh and Johor in the seventeenth and eighteenth centuries (*source*: Tate, 1971, 224).

albeit at times with difficulty, on the trading networks of Malacca and it undertook military incursions into Kedah and Perak. Its rival, Johor, exhibited similar characteristics at this time. Many of the remnants of the Malacca court settled there after the fall of the city and the economy of the state was based very much on the entrepot model with economic and political interests across and along the Straits.

THE STRAITS IN THE SEVENTEENTH AND EIGHTEENTH CENTURIES

It is perhaps tempting, especially given the nature of the surviving sources, to view events in the Straits in this period through the eyes of the increasing numbers of Dutch and English adventurers entering into the region. The volume of materials dealing with the activities of English 'country' traders and the Dutch United East India Company dwarfs that available from indigenous sources. Much of the non-European history of the Straits is thereby obscured. But it is important to remember that existing economic and political trends remained despite the impact of Europeans. Trading networks and relationships and traditional Malay–Muslim culture continued to develop and evolve during these years; indeed, without these, the European invaders would not have reaped much economic benefit from their incursions.

In 1602, a group of independent Dutch trading companies were amalgamated into the Dutch United East India Company (the *Vereenigde Oostandische Compagnie* or VOC) in order to facilitate, expand and ultimately control the lucrative trade of the Dutch, the premier European overseas traders at the time, with the East. It was a trade still dominated by spices, but increasingly based on other commodities—textiles, precious metals, drugs and perfumes. Two years earlier, their rivals, the English, were granted a royal charter to establish the East India Company with similar objectives, presaging an intense and sometimes bloody rivalry between the two nations for dominance in the East. Under the terms of a government agreement, the two were supposed to share the Eastern trade but, in the event, the Dutch easily dominated. After instituting the murder of a group of English traders at Ambon in 1623, the English virtually withdrew from the Far East to concentrate efforts on their Indian trade and territories.

Like the Portuguese they supplanted, the objectives of the Dutch were mercantile—to control the production, shipping and marketing of the key commodities of the region. Such a potentially lucrative task required a number of preconditions. A strong naval presence, at least initially, was important in securing strongpoints. As early as 1599, in order to secure the sites of spice production, they had seized Ambon and the Bandas, thereby controlling the precious trade in cloves and nutmeg, highly sought after on the European market. Through a series of military conquests and alliances, the Dutch sought to extend their stranglehold on production. To secure the shipping lanes, they began to take a series of strongpoints along the spice routes. Java was their first focus.

The west Javanese Sultanate of Bantam played an important part in the pepper trade of southern Sumatra and held a strategic position guarding the Sunda Straits. A powerful Muslim centre, it managed to resist Dutch pressures until the end of the seventeenth century. In 1619, the Dutch decided to set up a major trading factory close by at Batavia, on Bantamese territory, and the small port soon became the major centre for processing spices and for gathering Dutch merchants together before making the journey back to Holland in convoy. Relentlessly pursuing their monopolistic goals, the Dutch continued to combat Portuguese influence. The most important Portuguese outpost—Malacca itself—fell to the Dutch in January 1641. For the new governor-general, Anthony van Dieman, the taking of Malacca cemented the power of the Dutch in the region. 'The capture of Malacca,' he wrote, 'will excel all the conquests of the East Indian Company in India' (Lewis, 1995, 15). By securing two key ports—Malacca at the heart of the Straits and Batavia, close to the Sunda Straits and the kingdoms of Java—the Dutch now had a potential stranglehold on the trade of the Straits region.

The military successes of the Dutch did not lead to their total dominance of the economic and political life of the Straits. Like the Portuguese before them, they quickly found that traditional trading patterns and networks could prove difficult to tap. They also overestimated the importance of territorial control. As Lewis (1995) has shown, the seizure of Malacca did not produce the quick profits that the VOC had anticipated. As Figure 6.2 shows, adapted from data collated by Lewis, the trading performance of the city was disappointing to say the least. Whilst Batavia was more economically successful, what Lewis' analysis suggests was the continued importance of indigenous trade and traders in the region. Certainly the increased economic and strategic importance of Batavia did perhaps lead to a shift in the economic centre of gravity to the southern end of the Straits, and some Dutch trade did use the Sunda rather than the Malacca straits (Bruijn, 1980). But existing trading patterns were far from fatally disrupted by the arrival of the Dutch.

Alongside the Dutch, and perhaps benefiting from their presence, were a number of other states who continued to act as feeder ports and supply points into the wider trading circuits. The Dutch strength in the international market was critical to the success of this network of smaller ports. Only the Dutch had the naval, shipping and financial strength to trade on the European market. But the collection of the products needed for that trade—products still, as before, drawn from the local, China and India market, depended on the continued success of the traditional port towns and cities of the region. It may be argued that, at least for the first 150 years of their influence, the Dutch very assiduously controlled the mechanisms (and major profits) of a machine which had long been operative in the region, a machine whose networks, boundaries and transactions stretched along and across the Straits. It was not until the latter part of the eighteenth century that this pattern was changed as the Dutch intervened much more purposefully in the production of new, tropical commodity crops.

Traditional states such as Aceh, whilst losing some of their power to the Dutch, nevertheless remained important. Certainly the Dutch began to encroach on the two major products underpinning Aceh's growth—Sumatran pepper and tin from Kedah and Siak. But, despite a series of conflicts in the late seventeenth century, the Dutch were unable—or unwilling—to conquer Aceh. Eventually the Dutch drive to secure monopolies of the pepper trade led to the decline of Aceh as a powerful, independent kingdom, but not to the disappearance of those trading networks that had animated its economy. The Dutch continued to rely on these indigenous systems for their major supplies.

Muslim trading networks continued to be very important in the Straits, especially

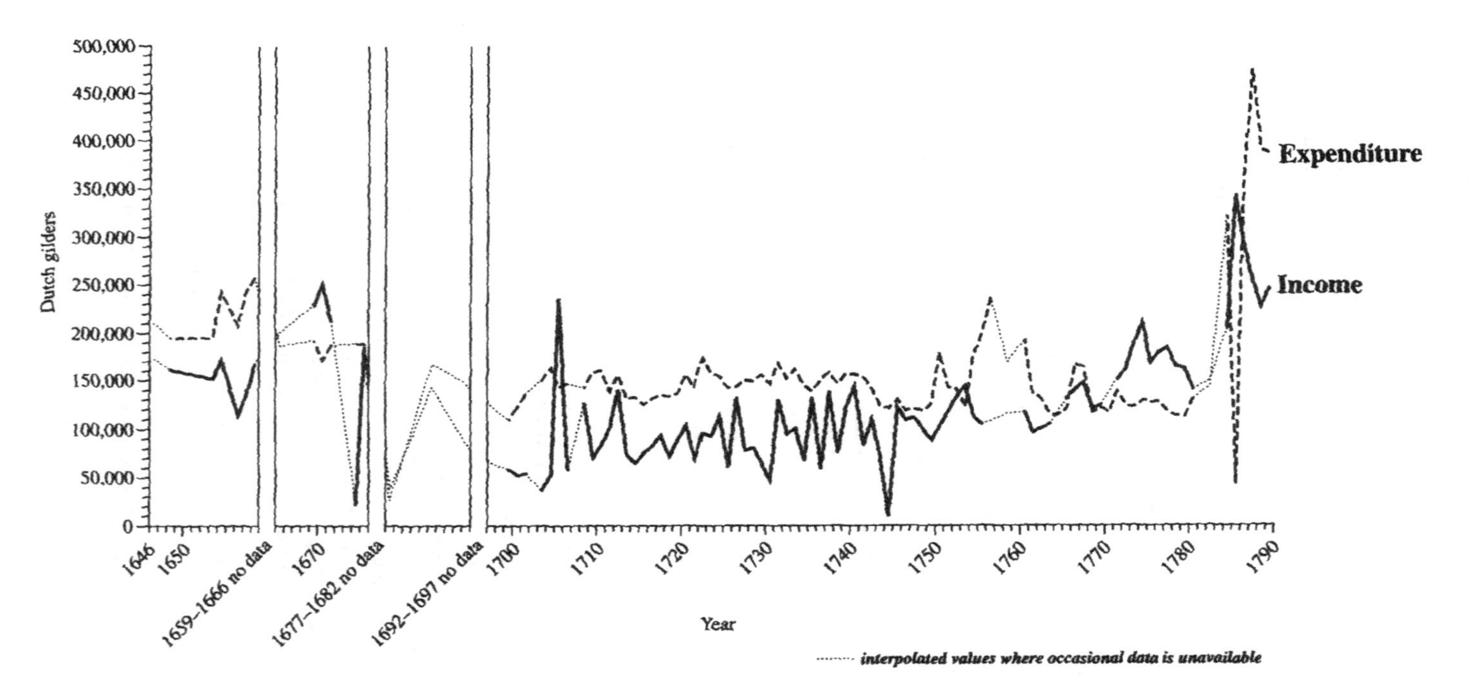

Figure 6.2 Income and expenditure of the Dutch at Malacca, 1646–1790 (*source*: Lewis, 1995, appendix)

in the smaller, sub-regional patterns that fed into the wider international trade. Thus the trade with south India was dominated by Muslim traders, at least until the power of the British East India Company was consolidated in the late seventeenth century. The *orang laut*, so important as local traders and collectors continued to be significant on both sides of the Straits. Indeed their activity continued independently of external political changes and may well have grown during the years of Malacca's decline. They certainly continued to play an important part in the activity of the still powerful Johor–Riau kingdom to the south.

Amongst a range of indigenous groups who played an important role in the economic life of the Straits, by far the most important were the Bugis (Pelras, 1996). The Bugis had been migrating in considerable numbers from the end of the seventeenth century into the Straits region. Forced out of their homelands in Sulawesi by Dutch intervention, they settled widely on the coast and islands of the southern end of the Straits. They were highly-skilled sailors and navigators—arguably the best in the region—as well as being inveterate traders. Their good fighting skills made the Bugis formidable allies for Malay leaders and the Dutch alike. By the 1730s, through a series of local battles and alliances, the Bugis had consolidated their hold in the Johor–Riau kingdom, largely supplanting traditional elites and increasingly dominating those local trading networks once the domain of the *orang laut*. Their influence spread throughout the Malay side of the Straits, with power bases in Selangor and Linggi as well as in Johor. In 1757, Bugis forces almost succeeded in taking Malacca itself, much to the embarrassment of the Dutch.

By the end of the eighteenth century, the position of the traditional port city was under threat. Given the relative paucity of their land resources, traditional states had always relied heavily on trade, both short- and long-distance, for their wealth, status and power. The kingdom of Srivijaya epitomised this, as did, at a later date, Malacca, Aceh and Johor. The arrival of European powers did not at first fatally alter the underlying economic role of these states. Certainly the reins of power were in different hands but, in many ways, the Portuguese, English and Dutch altered the apex of a system which, at base, remained remarkably persistent. It is important to emphasise how both trading and political systems in this period extended over both sides of the Straits. Srivijaya, Malacca, Johor and Aceh had territories, trading links and political relationships equally on the Sumatran and Malay sides of the sea. The Straits were a force for unity rather than division.

By the century's end that position was beginning to alter as the nature and intensity of European intervention changed. A shift from mercantile, trading interests to more concerted industrial, production-orientated interests led to important changes in the character of political and economic life. The actions of the Dutch in the seventeenth century in extending their control to both the marketing and production of spices in the Moluccas presaged a new, more intense form of intervention further west. By the latter part of the eighteenth century that intervention was making itself felt in the Straits.

7

COLONIAL INTENSIFICATION AND THE RISE OF SINGAPORE, 1780–1930

By the end of the eighteenth century, the character and intensity of European intervention in the Straits had begun to change. Whilst the economy, society and cultures of the peoples of the region had long been open to external influences from China, India, the Middle East and Europe, by the 1780s it was the influence of the latter which came increasingly to dominate. Changing patterns of economic development in Europe stimulated by the Industrial Revolution, made the economic and political colonialism of the European powers, especially Britain and Holland, the dominating force in the Straits.

SETTLEMENT AND ECONOMY

Perhaps the most striking feature of the historical geography of the Straits in this period is the strong expansion and consolidation of European economic and political power which was manifest in many aspects of the life of the region. But we should nevertheless not underestimate the continued importance of local societies and polities which can easily be overshadowed by the greater intensity of British and Dutch influences through the Straits. The wider networks of social and economic interchange in the Straits remained underpinned by the small-scale, short-distance, everyday trading activities of the *orang laut*, Bugis and Malays in the region. Malay and Chinese traders continued to oversee sailing and trading in the estuaries and islands around Malacca, Aceh, Perak and the Johor-Riau region even though, at the larger scale, it is the degree of European intervention which is most apparent from the records.

The development of Johor-Riau illustrates this well. The strength of traditional Malay sultanates in the Straits had always been a reflection of the immense sailing and trading skills of the peoples of the islands, and the rise of Johor-Riau as an important trading port and market at the southern end of the Straits reflected this native skill. Married to the entrepreneurial skills of Chinese traders and Malay elites, the Bugis' sailing skills made a formidable and powerful combination (Pelras, 1996). As we noted in Chapter 6, the Bugis, archetypal sailor-traders in the region, had been able to seize power in Johor-Riau such that by the 1760s Bugis domination of the state was secure. Riau became an important regional trading centre, but its staples were no longer the spices and forest products that had attracted foreign interest in the past. The development of gambier plantations (gambier was an important dyestuff used increasingly in industrially-produced textiles), coupled with the continued importance of tin and pepper, both now produced on an increasingly large scale, made Johor-Riau an important centre for what came to be termed the 'country trade'. The lack of tolls and taxes, coupled with a profitable sideline in trading in opium for the Chinese market, further added to the importance of the state.

The development of the country trade was to be a key element in both local and regional prosperity. Its growth reflected the increasing importance of the China market to western, especially English, industrial producers and consumers. The English East India Company, by virtue of its royal charter, exercised a monopoly on the key trade between Britain, India and China. In theory, only the ships of the company could trade directly with the Chinese market, ensuring thereby a profitable monopoly to the company. However, as the China trade developed, demand for Chinese products—especially higher value goods such as tea—expanded in Europe. But the company always required products which could be exchanged for these goods because, from the late eighteenth century, a marked imbalance in trade between the two economic spheres began to emerge. To save valuable currency and gold from having to be sent to China, a process which would quickly have drained treasury away from the company, attention was turned to sourcing products in Southeast Asia which might serve as exchange goods on the China market. Both the Dutch and British trading companies were therefore happy to encourage the new so-called country traders, usually Europeans, to scour the archipelago for products to be traded on to the China market. Such traders were usually paid in kind or credit notes redeemable through the company.

This country trade was important in a number of respects. First, whilst the China trade had long been important in the Straits, it now both intensified and came increasingly under European control. Certainly, local groups continued to play a vital role in the sourcing and collection of produce, but its integration into the China market fell to a new group of middlemen, now increasingly European rather than Malay or Chinese. Second, these largely English and Dutch country traders were the agents who facilitated the integration of short- and long-distance trading networks in the Straits. They bought from local producers and collectors and sold on these products to the long-distance Dutch and British trading companies in exchange for the new industrial products (textiles, metal goods, mass-produced ceramics) which were then sold back onto local markets. Without the development of country trading, that circuit of trade and profit could not have been completed (Lewis, 1970).

The role of the country traders as circulators of both goods and capital became a key element in the continued activity and prosperity of the ports of the region. Those that could adapt swiftest to the new demands of trading prospered. Johor was amongst the ports that did well. With its Bugis masters highly adept in both the legal and illicit trade in tin, opium, pepper and gambier, it prospered and soon attracted once more the attention of the Dutch. In 1784, anxious for a slice of the lucrative trade in the southern Straits, they seized Riau and reduced the power of the Bugis. That seizure proved fatal for the port for, as happened elsewhere, the Dutch could replicate neither the conditions of prosperity nor the trading skills of Riau, and the port went into decline (Trocki, 1979).

The rising importance of the China and country trade to British interests was also reflected in a search by the East India Company for a new British base in the region. With the Dutch holding Malacca at the northern entrance to the Straits and Batavia, close to the southern entrance, the board of the company had commissioned reports on a number of potential sites. Their motives were both strategic and commercial. A British port in the Straits could serve for reprovisioning of both military and trading ships at a convenient mid-way point on the India–China route.

Furthermore, any port site that had the potential to tap into a valuable hinterland might be useful for collecting goods for the country trade, thereby allowing the company some input into this lucrative trade. In 1782, the decayed Aceh sultanate was the first site of an abortive attempt to establish a company base. Two years later, an attempt at settling company facilities at Riau was also unsuccessful. Finally, in 1786, Francis Light, a country trader working on behalf of the company, established a settlement on Prince of Wales Island, later to become Penang, and founded the port of Georgetown as its capital. The island appears to have been ultimately selected because its owner, Sultan Abdullah of Kedah, was prepared to lease the island to the British.

As Andaya and Andaya note (1988, 107), 'for Malays, the establishment of Penang transformed the [East India Company] into a territorial power with an obvious stake in the security of the area'. But for the company, Penang was far from the great success that had been anticipated. Once more, European powers had assumed that geographical location, by itself, was enough to guarantee success. The trade of the island did of course increase, though that increase was steady rather than spectacular. But the costs of cultivating spices and pepper were high, primarily because of the high costs of labour. The returns for those goods were, in any case, declining on the European market (Cowan, 1950, 6–7). In addition, hopes that the port might become a suitable naval base and shipbuilding centre were dashed by two factors. First, the high costs of both infrastructure and materials (especially timber) made the economics of shipbuilding prohibitive. Second, following the Battle of Trafalgar in 1805, British fears of a large French naval presence in the East receded, and both company and government interest in the strategic importance of Penang waned.

The relative lack of success of Penang showed once more that a good strategic position in the Straits was not sufficient to guarantee commercial and political success. That success continued to depend on the ability to integrate local, regional and international trading networks. Monopoly positions, such as those both the Dutch and British often sought to impose, simply dissuaded traders from coming to the ports. Local traders went elsewhere—to Johor, to Riau or to Aceh—whilst the company ships themselves often had insufficient cargo room on their return from China to even go to the trouble of calling at Penang. The products of the island in which such high hopes (and capital) had been invested often languished unsold in London warehouses or, more frequently, remained in store on the island.

Penang was also affected by wider political changes in the region. Between 1810 and 1816, as a consequence of the French occupation of Holland, the Dutch Crown which had fled to Britain, granted the British government temporary control of its possessions in the East. Malacca, southern Sumatra, Batavia and most of Java came under British control in those years. Trade which had once come to Penang to avoid high taxes at Dutch ports (which, like their British counterparts, sought to monopolise shipping and trade), now returned to those ports under British control. For Cowan (1950, 6), 'the recession in the trade of the island seems to have been the direct result of the British occupation of Java and the disturbed state of Achin in the North of Sumatra'.

Increasingly by-passed by the India–China trade, the high hopes Francis Light had placed in the island settlement had evaporated by the 1820s. Nor did the return of the temporary British possessions to the Dutch after 1816 help matters, for the British government, anxious to keep the Dutch powerful in Europe, were unwilling to countenance too strong an opposition to the renewed commercial vigour of the Dutch in the East. Malacca, whose fortifications had been torn down during the British occupation in the hope of reducing still further its rivalry to Penang, was handed back to the Dutch in 1818, further damaging Penang's hopes of becoming the prime port of the region. One year later, the English adventurer and administrator, Thomas Stamford Raffles took over the Malay settlement of Singapore in the hope of making it the primary port in the region.

THE BRITISH AND DUTCH IN THE STRAITS

By the close of the eighteenth century, the character and intensity of European involvement in the Straits was changing inexorably. Central to those changes were important shifts in the nature of economic intervention. Control of trade was widened through greater military and naval intervention and, crucially, both powers began to involve themselves directly in the production process. No longer content with simply marketing the products of the region, growing demand in Europe stimulated the growth of European investment in the production process in a range of mineral and agricultural fields. What of the traditional powers in the region at this time of fundamental change? Three processes were altering the status and role of traditional polities on both sides of the Straits.

First, it is clear that patterns of trade were increasingly shifting into European hands which were better organised, had greater access to capital and, through the major trading companies, were able to exercise an increasing stranglehold on trading networks. Whilst the success of European trade still depended on cooperation with good local trading groups (the case of Penang illustrates this well), the ultimate control and manipulation of those networks (and the real profits) were increasingly in European or Chinese rather than Malay or Sumatran hands. Thus, for example, the growing importance of country traders heralded an inexorable shift in the balance of trading and economic power from indigenous to foreign hands. The extent to which trading networks at local and sub-regional scales could remain controlled by indigenous traders was increasingly compromised by the greater economic muscle and organisation of the Europeans.

Second, it is evident that both the Dutch and British, not content with the profits to be gained from trade, began, from the late eighteenth century, to concern themselves with direct control of the production process. The Dutch were the first. They had, after all, done precisely that through their plantations in the Spice Islands from the late seventeenth century onwards. That experience provided them with a model they could apply elsewhere. As the nineteenth century progressed their interest in, and involvement with, the production of pepper, tea, coffee, tobacco and rubber became increasingly important to the survival of the maritime cities of the region. On Java, the development of the Culture System from the 1830s heralded the forced production of a range of cash commodities which were traded through Dutch ships and merchants. On the Sumatran side of the Straits, as we noted in Chapter 5, the production of tobacco in the Deli region brought about the wholesale transformation of large areas of land into the plantation system.

Paradoxically, then, control of land and its resources became crucial to the survival of the quintessentially maritime cities of the Straits. For traditional peoples of the region this marked a new and unfamiliar turn for they had always looked to the sea for their livelihood and wealth. For many, as Trocki (1979, 205) notes, land had been traditionally regarded almost with disdain: 'as port and sea chiefs, they saw the land as a place from which one gained wealth, not as a place in which one invested resources. The business of governing a piece of land was not the proper occupation of a true maritime ruler'. That failure to embrace the need to become involved in the land-based production process was a costly error.

Third, increased European intervention in the region signalled much greater 'policing' of the Straits by foreign powers. Part of that policing was reflected in attempts to suppress piracy. Now, our definition of piracy has been furnished largely through European sources. What was regarded by traditional traders and *orang laut* as the legitimate levying of taxes, 'gifts' and exchanges, sometimes legitimated by force, sometimes by custom, was to the European, piracy. Increasingly, something which had been part and parcel of the maritime life of the region, and an important source of revenue to indigenous peoples, faced suppression by the increasingly interventionist European powers. The clash between indigenous notions of traditional exchange and European notions of free trade could hardly have been more starkly drawn than over the issue of piracy. Anything which interfered with the free flow of traffic and goods—or rather the free flow of Dutch and British traffic and goods—was a target for antipiracy actions. But maritime 'peace' was for many Straits peoples, the 'peace of oblivion' as the English and Dutch sought to wipe out native commerce altogether (Trocki, 1979, 208).

These three inter-related processes were to place important constraints on native maritime cities. Aceh, long in decline from its hey-day in the seventeenth century, was moribund by the late eighteenth century. As Tate (1971, 226) noted, 'its rulers no longer lorded over the northern waters of the Straits or dominated the peninsula states opposite'. By the early nineteenth century, the British had also begun seeking treaty agreements with the sultanate as part of a policy of establishing a base in the northern Straits. Further south along the Sumatran coast, the former powers of Siak and Jambi, once important nodes in the Achinese empire, were also in decline, quite unable to attract the trade of the Straits, and unable to match the increased military and economic strength of the Dutch. By the mid-nineteenth century they, like Aceh, had succumbed to Dutch control.

Parts of southern Sumatra had been fought over by the powerful Johor-Riau empire, the last Malay polity to exercise power over both sides of the Straits. Riau was a very important centre of the country trade in the mid-eighteenth century, and the arrival and consolidation of Bugis power made it a key trading centre in the region, more powerful than any other (Tarling, 1962, 3–12). But the arrival of the Dutch, who took Riau in 1784, weakened the Johor empire and the founding of Penang further weakened Johor. The weakening was not fatal, for Johor was to revive in the mid-nineteenth century with the growth of Singapore, but, as Trocki points out (1979, 41), 'between 1819 and 1850, Johore was no more than a geographical expression'. Further north, Malacca was almost moribund as a port and trading centre. Her fortress had been pulled down by the British and most of her mineral deposits were exhausted: 'her commercial role had declined to a shadow of its former greatness' (Sandhu & Wheatley, 1983, 518).

It was British and Dutch intervention that was in large part responsible for the parlous state of affairs amongst local powers in the region, and external power struggles were responsible for a European treaty which, for the first time in the region's history, led to the formal separation of the Malay and Sumatran coasts of the Straits—the Anglo-Dutch Treaty of 1824. As we have noted earlier, with the onset of the European Napoleonic Wars, Java and Sumatra were transferred to British hands in 1796. With Malacca and Penang controlled by the British and Johor-Riau in decline, Britain became, almost by accident, the paramount power in the Straits at the start of the nineteenth century. With the ending of the Napoleonic wars in 1815, the problem of Dutch–British relations in the East was a significant one.

As Tarling has shown (1962), there were a series of important debates within British political and commercial circles over the terms of territorial returns to the Dutch. Should Britain, for example, keep the ports of Bangka (on Java, abutting the Sunda Strait) and Riau in order to keep an eye on any Dutch economic revival? Should Malacca be returned to the Dutch, a move which might damage the commercial growth of nearby Penang? Raffles had secured treaty rights with Aceh in 1819: should these be strengthened through territorial control? It was even mooted that newly-founded Singapore might be 'given back' to the Dutch despite its growing trade role (Tarling, 1962, 133). In the period preceding the Treaty of 1824, the Straits had clearly evolved all the features of a dependent, quasi-colonial region—its seas and boundaries argued over in the courts and offices of London and Amsterdam. The key factor determining the outcome of these agreements was European politics not

Southeast Asian realities. As Tarling argued (1993, 27), 'an effectively independent and friendly Dutch state was important to the British in Europe . . . a Dutch empire overseas was the price to be paid'.

Under the terms of the treaty, Singapore and Malacca were retained by the British, Bencoolen and British dependencies in west Sumatra were ceded to the Dutch, and two geopolitical boundaries were drawn creating two spheres of influence. The British agreed not to establish further outposts south of Singapore and any British influence in Java was ended. A second line was drawn plumb down the Straits of Malacca. To the west of this imaginary line, the Dutch were to be the governing power, to the east, the British. Thus, crucially, the Malay Peninsula and Sumatra were separated politically and economically by a European treaty. The Treaty of 1824, signed in London, effectively divided regions which culturally, economically and politically had far more in common with each other than with the powers who now claimed authority over them. That division was to be enshrined in both the colonial and post-colonial settlements of the twentieth century.

THE EMERGENCE OF SINGAPORE

It was Sir Stamford Raffles, writing in 1819, who first articulated his colonial vision of how the newly-founded settlement of Singapore, at the tip of the Malay Peninsula, would develop:

> You may take my word for it, this is by far the most important station in the East, and as far as naval superiority and commercial interests are concerned, of much more value than whole continents of territory. . . . It would be difficult to name a place on the face of the globe with brighter prospects.
>
> (Quoted in Tate (1979, 149.)

Within two decades, such predictions were proving correct as the port developed a range of trading networks and facilities to service the growing British colonial interests in Southeast Asia. By the end of the nineteenth century, its trading role far outstripped the geographical confines of the Straits of Malacca and the South China Sea. Indeed, by 1930 it was no longer a primarily 'colonial' port, dependent on trade with Britain and British imperial possessions. Its interests had become truly global. As its activity expanded, the merchant community, agency houses and shippers of Singapore traded globally, from the new markets of Australia to the growing industrial economies of the United States and Japan.

The study of Singapore's growth is important not only in itself, but also for what it reveals about the nature, patterns and development of global and regional trade in the nineteenth and early twentieth century. Like many of the maritime cities of the Straits, it grew through a combination of geographical position, local skill and acumen, and the impact of European economic and political intervention. Its foundation came at a time when important underlying shifts were occurring in global trade: the important early links between colonialism and economic growth, the increased demand for key export commodities (tin, rubber, hydrocarbons) on the world market and the dramatic impact of new technologies on the nature and geography of maritime trade.

The growth of Singapore to its position not only as the key port of the Straits region by the late nineteenth century but also to a position as a major global port is perhaps the most exciting aspect of economic change in the Straits in this period. The dominance of Singapore in the early twentieth century rivals that of Aceh in the sixteenth or Malacca in the fifteenth. What factors underlay this growth? We would suggest that three principal factors need to be examined. The first concerns the relationship between port activity and colonialism. To what extent was the growth of trade linked to colonial interests, whether through the long-distance trade with Britain and India, or short-distance trade with other parts of Southeast Asia? And, linked to this question, how far was the emergence of Singapore as a major global port in the first decades of the twentieth century a reflection of its ability to diversify its trading links away from traditional colonial ties?

Second, technological change, notably through advances in shipping technology, clearly impacted on the nature and geography of Singapore's trade. Improvements to sailing ships, and the development of steamships from the mid-nineteenth century, were to have important consequences for the nature, quantities and geographical patterns of imports and exports, as we have emphasised in Chapter 5. A third important factor is the role of capital. The inter-relationships between local and European capital, especially in the development of the major shipping lines operating from and through Singapore, and the links between private and government capital in the development of port facilities, may also have contributed to the changing role and importance of the port and its hinterland.

When Raffles decided to establish a free port on the island of Singapore, there was already a Malay settlement in place which had trading relations in the region, particularly with the Bugis traders whose role in the coasting and long-distance trade between the Mollucas, southern Borneo and the Malay Peninsula was well established. The early growth of Singapore thus reflected in part the geographical advantages of the position it occupied at the southern end of the Straits of Malacca, and the network of indigenous traders and merchants already active in the region: its success cannot be ascribed solely to the colonial influence.

As with Penang, founded some 40 years earlier, in establishing the settlement, British traders sought a site from which to tap into the trading networks of the region and also to provide a location from which to monitor and, possibly, rival Dutch commercial might. With Penang and Malacca, now ceded to the British by the Treaty of 1824, Singapore formed the third of the so-called Straits Settlements, the first significant British footholds in Southeast Asia. Singapore became the capital of this administrative region in 1837. Initially governed as part of the Indian Colonial Service, the Straits Settlements finally became a Crown Colony under the authority of the Colonial Office in 1867 (McIntyre, 1967, 69). With the consolidation of these three settlements, a pattern presaged in the Treaty of 1824—British influence on the peninsular, Dutch control in Java and Sumatra—was set in train.

Singapore itself was first established as a British settlement in 1819, with the whole of Singapore Island being ceded by the Sultan of Johore five years later. Under the direction of Raffles, it was from the outset established as a free port. Unlike a number of other ports in the region where taxes, monopolies and various 'gifts' were extracted by local potentates, Raffles and later administrators insisted on free and untramelled trade. Its geographical position meant that it rapidly became 'a natural half-way house and transhipment centre for traders from China and the West' (Tate, 1979, 150), although geographical position alone is not sufficient to explain its rise. Penang, too, was in a good geographical position and a free port, yet singularly failed to realise the high hopes invested in its establishment.

Singapore grew spectacularly fast but was always vulnerable to cycles of boom and bust. Population grew from around 16,000 in 1827 to just over 80,000 by 1860 (Turnbull, 1989, 36), with about 65% of that population comprising migrants from southern China. With a sprinkling of European traders and administrators, a growing group of so-called Straits Chinese (later know as Baba Nonya), Indians, Malays and Bugis, it was always a mixed, cosmopolitan and dynamic city. Its trade grew considerably in the early years. A total trade of around $11 million had grown to nearly $17 million by 1833 and $55 million by the early 1860s (Wong Lin Ken, 1960, 51).

Trading conditions however could change rapidly. As Turnbull suggests (1989, 42), 'almost entirely dependent on a fickle entrepot trade, Singapore's position in the mid-nineteenth century was delicate and her trade was subject to violent fluctuations and to years of uncertainty and depression'. The point is worth underlining—Singapore's pre-eminence in the region was neither certain nor continued. It was not, furthermore, a pre-eminence which relied solely on the establishment and consolidation of colonial ties. Much of its trade, at least in the first half of the nineteenth century was regional rather than international: Wong Lin Ken (1960, 50) estimated that as late as the 1860s trade between Singapore and the west rarely rose above 25% of total trade. Changes in demand, the transfer of much of the important Bugis trade to the new port of Macassar (today's Ujung Pandang) on Sulawesi from the 1870s, and competition with the ports (and protectionism) of the Dutch East Indies, especially Java, often created difficult conditions for the fledgling city, and required flexibility and adaptation to changing economic circumstances for the merchant and banking community.

TRADING REGIMES IN THE STRAITS

The trading regimes of ports in the region during the nineteenth and early twentieth century were inevitably diverse but in essence can be characterised under four related groups: the coasting trade, the country trade, entrepot activity and, finally, long-distance trade. Whilst the latter is frequently the most striking in analyses of trading patterns, the other three are of at least equal significance in the development of Singapore. What then were these different regimes?

The coasting trade was integral to the development of the maritime economies of Southeast Asia. Ships involved in the coasting trade focused on short-distance trade, hopping from the sanctuary of one small port to another, from one estuary to another, carrying small, locally-traded cargoes ranging from rice to cotton goods, to woods and rattans. A network of ships and local carriers thus connected the often isolated economies of the northwest and southern coasts of Borneo, the Outer Islands of the East Indies, the coasts of the Straits of Malacca and linked those economies with the major ports such as Malacca, Batavia, Palembang and, later, Singapore. Certain ethnic groups dominated this trade—the Bugis traders from the island of Sulawesi were especially important. The arrival of their native ships or *praus* in the larger ports of the archipelago heralded a buzz of trading activity for local and foreign merchants alike. They provided the important feeder connections between the indigenous economies and products of coast and up-country with the wider networks of international trade.

The second category—the country trade—was to some extent linked to the coasting economy though its structure was rather different. English merchants and traders were especially active in the country trade which developed, as we have seen, largely as a response to the growing importance of the wider trade between Europe, India and China. Given that the ports of Southeast Asia—and especially Singapore—were stopping-off points for the Europe–China trade, English 'country traders' began to scour the archipelago for a range of products which they could trade on the China market in exchange for the sought-after Chinese exports. The country trade thus grew and, linking in with the native coasting economy, provided a further means of channelling indigenous products onto the international market. In this way the circuits of indigenous and foreign trade, shipping and capital began to interact. The English 'country trader' would advance capital to Malay or Chinese merchants to allow them to collect indigenous goods which, in turn, would be sent down-river to enter into the regional and international market, access to which was controlled by the traders and the larger shipping lines.

Entrepot activity was destined to become a third important arm of the trade of Singapore. Its growth was also linked with the coasting and country regimes described above and, as with them, its development increasingly brought together different circuits of capital and trade. The increasingly important position of Singapore as both a regional centre and as a stopping-off point for international trade between Europe, India and China led to its use for both the storage and processing of a range of commodities. Thus many of the products brought to Singapore through the coasting trade required sorting, classifying and packaging prior to their onward transit. Products ranging from jungle rubber and spices to precious woods and minerals usually required some form of processing and Singapore developed such facilities early on.

This entrepot activity—the collection, processing and storage of commodities prior to onward transit—became one of the keys to the commercial success of the port. It is also worth emphasising how such activity increased the interaction between indigenous and foreign circuits of capital and trade. Whilst the products flowing into the port required the action of local traders and coasters, the costs of storage and processing, which could be considerable, were frequently borne by European and Chinese traders and merchant houses. By the mid-nineteenth century, many such houses had developed closer links with indigenous traders and coasters by advancing credit on the strength of the products that would flow into their warehouses and godowns. Once collected, these merchant houses were pivotal in channelling goods onto the major shipping lines. By the end of the nineteenth century, the nexus of interests between international shipper, Singapore agency house, the merchant banks and the indigenous producer was becoming increasingly close.

The fourth category—long-distance trade—represented in many ways the most visible and spectacular aspect of the growth of Singapore. From the mid-nineteenth century, important British shipping interests such as P and O, Holts 'Blue Funnel' lines, the French line, *Messageries Imperiales* and the Dutch state line, *Koninklijke Paketvaart Mittshappij* (KPM), were playing a major role in the shipping and trading enterprises of both Singapore and the wider region. Their activity, however, was built on a network of feeder lines, coasting links and agency and merchant houses which fed products into the networks of trade and provided important sources of revenue and profit for the companies. Their activity was also, by the end of the century, hedged around with a whole raft of restrictions usually incorporated within complex arrangements known as conferences which sought to divide shipping trade between the major companies

TRADING PATTERNS THROUGH SINGAPORE

The sources for reconstructing the trading patterns of Singapore in this period are varied both in nature and value. An evaluation of imports and exports, usually on an annual basis, is provided by the trading returns published as Blue Books for all of the Straits Settlements. The amount of detail in these returns varies. Generally, they provide quantitative estimates of imports and exports broken down by product, origins and destination. Monetary values in Straits dollars together with values by weight are also provided. In addition, some of the returns list added information. The number and origins of ships cleared in the ports, a breakdown of local and 'foreign' vessels, and the balance between steam and sail ships may also be available. The problems of using such official data have been described in some detail elsewhere for the Borneo trade (Cleary, 1996, 1997); much the same precautions need to be taken in examining the Singapore trade. However, creating a data base from these sources, whilst problematic because of temporal and spatial variations in the data, provides the basis from which to examine key changes in trading patterns.

In examining the ways in which the trade of Singapore evolved between 1860 and 1941, the work of Wong Lin Ken (1960) is of particular importance. In a pioneering paper published in 1959, he analysed the evolution of Singapore's trade in the period

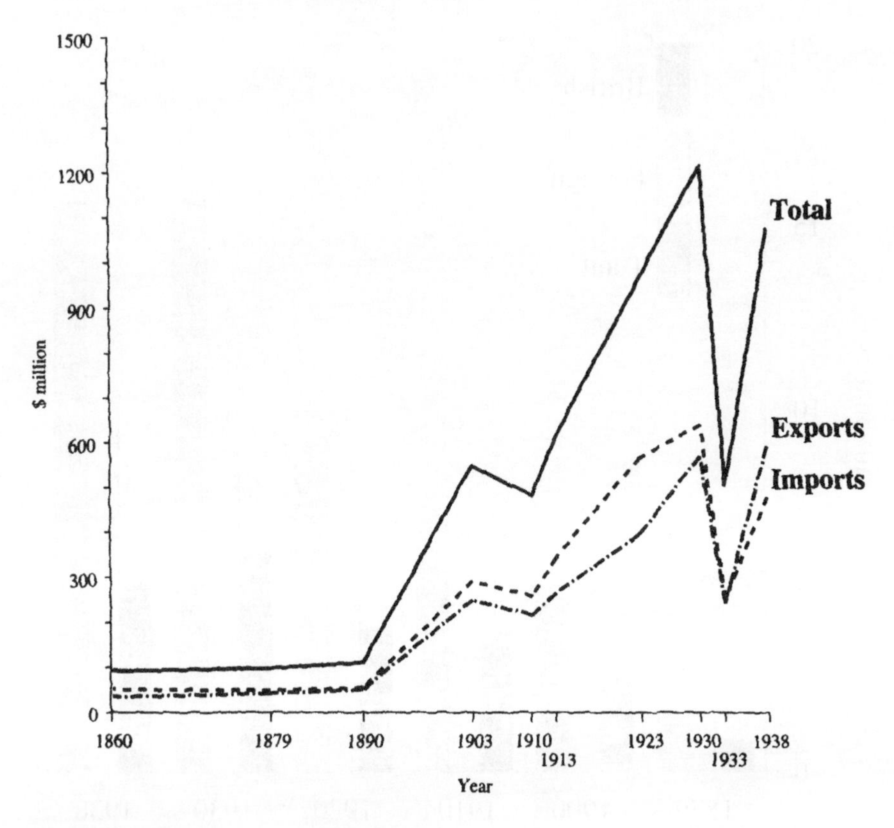

Figure 7.1 Imports and exports through Singapore, 1879–1938 (*source*: Trade Returns).

1819–1869 making use of the range of trade returns available for that period. Subsequent work has developed the analysis for later periods, although his later work does not elucidate in the same detail some of the trends identified in the 1959 paper. The analysis here thus builds on the work of Professor Wong and seeks to extend some of his work into the later period and the importance of his work is acknowledged here.

Of prime importance is the analysis of changes in the volumes and direction of the trade of the city. Using the trade returns it is possible to provide broad estimates of changes in trade. Figure 7.1 shows how the combined imports and exports of Singapore grew between 1860 and 1938. Over that period the volume of trade passing through the port grew from just under Straits $42 million (all subsequent figures are in Straits dollars) to almost $600 million, a fourteenfold increase over the period. The broad picture of increasing trade does conceal periods of growth and retrenchment. Thus the spectacular growth in the first decade and a half of the twentieth century reflected, to a considerable degree, the growth of tin mining and rubber planting on the Malay Peninsula, whilst the retrenchment of the late 1920s

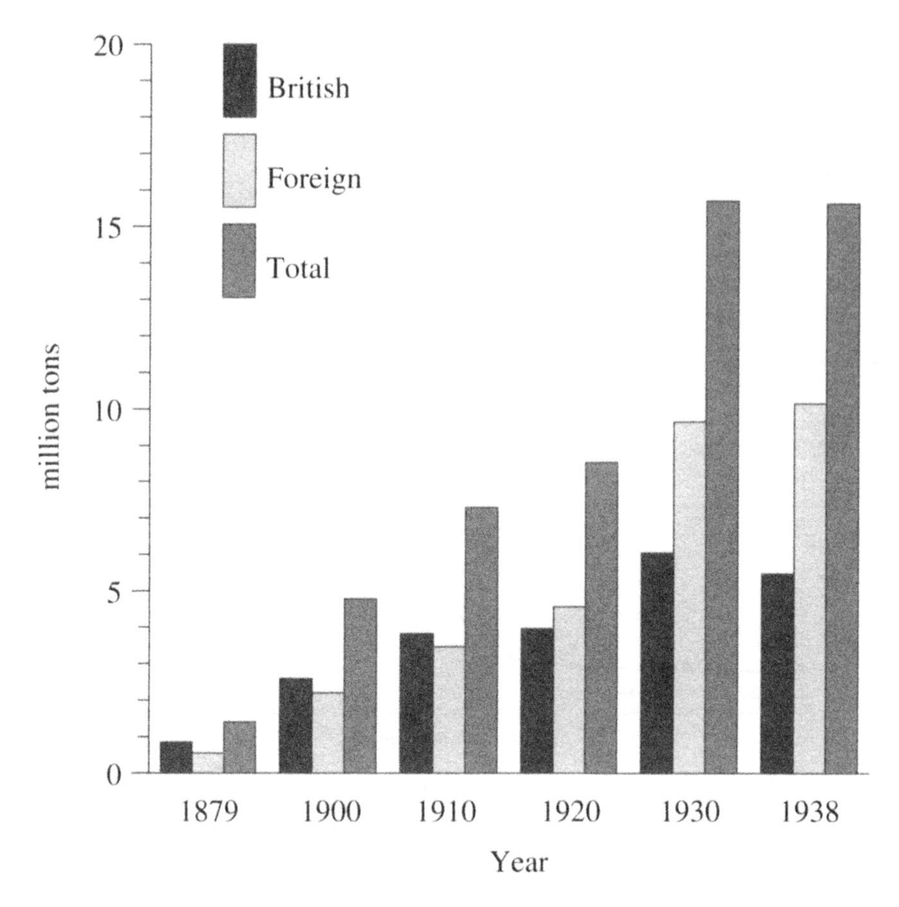

Figure 7.2 British and non-British ships through Singapore, 1879–1938 (*source*: Trade Returns).

and early 1930s were a consequence of declining global demand for these same primary commodities during the economic depression.

Underpinning this sizeable growth in trade were a number of other changes. Not surprisingly there were massive increases in the average tonnage of ships passing through the port. Figure 7.2 shows changes in the total gross tonnage of ships using the port between 1879 and 1938. The data also reveals the balance between British and foreign registered ships in the port; they underline the extent to which foreign registered ships were increasingly outstripping British boats by the end of the period under study. Figures on tonnage are, however, not always straightforward to use. For many of the smaller native vessels, which were a key part of the coasting trade, the figures are either missing or inaccurate, suggesting that the increases in tonnage are, if anything, underestimates of the true size of the trade.

These changes in gross shipping tonnages were accompanied by increases in the average tonnage of ships using the port. Although, again, the data is not wholly reliable, the average tonnage of non-native ships rose from around 779 tons net in 1879 to 1,386 tons net in 1910 and 1,737 tons net in 1930. By 1938, the average net tonnage of ships using the port stood at around 2,414. By the end of the 1930s, then, more ships were using the port, their total and average tonnage had greatly increased, and the range of registration ports of vessels in the harbour was very much wider than it had been some four decades earlier. By a range of quantitative measures, then, Singapore had moved into the first rank of global ports.

Accompanying this major transformation in trade were a range of important infrastructural changes in the port itself. Good berthing facilities, the availability of a range of bunkering services and the proximity of major financial, agency and insurance companies were essential to the development of Singapore as a global port. Up until the mid-1840s, most shipping using Singapore was concentrated in the old heart of the city, using berths along the Singapore River and the Collyer (or Boat) Quay. By mid-century, these facilities were becoming increasingly congested and the development of Keppel (or New) Harbour was begun. The administration of the colony, anxious to avoid any expenditure, refused to sanction works at its own expense and private capital was used to develop facilities. Thus in 1861 the New Harbour Dock Company was established and three years later its rival the Tanjong Pagar Dock Company began to develop facilities and reclaim land for new quays to service the growing volume of trade. By 1899, the latter company exercised a virtual monopoly on facilities and had absorbed the New Harbour as well as the Borneo Company and Jardine wharves.

By the turn of the century, however, facilities had still barely kept pace with growth. The face lines to many of the wharves were irregular and there were no railway links. In 1903, the government was persuaded, somewhat reluctantly, to abandon its laissez-faire principles, expropriate the Tanjong Pagar Dock Company and embark on a programme of modernisation. The main agent of that modernisation, the Singapore Harbour Board, was set up in 1913. As a result new wharves and road links were built and the giant Empire Dock was opened in 1917.

FROM COLONIAL TO GLOBAL PORT

It is clear from the preceding section that the volumes and value of goods shipped through Singapore experienced dramatic rises in the two decades either side of the turn of the century. By the early 1930s, Singapore was estimated to be the fifth or sixth most important port in the world. But the analysis of the volumes traded through the port gives only a partial picture of its development. In seeking to analyse how the port grew it is equally important to consider the nature of products traded and, most importantly, their origins and destination. If Singapore emerged as a truly global port, this should be fully reflected in its trading profile.

Fortunately the trading returns, coupled with data from individual shipping records, allow a detailed geographical and product profile to be drawn. Thus for most years it is possible to get a breakdown of imports and exports by country, together with data on the port of registration of foreign ships (but not, unfortunately, of so-called 'native craft') and the nature of products moving in and out of the port. Analysis of that data set can provide important clues about the changing nature of Singapore's trading role.

Given the strong colonial origins of the port, a first task is to establish the importance of colonial links in that trading profile. The relationship with Britain and with India was clearly a fundamental factor in the early growth of the port. That linkage was important in sourcing imports into the city as well as in directing the export trade. Thus in the first few decades of its development, when Singapore was governed through India, much of the trade of the city reflected that colonial connection. Textiles, manufactured goods, coal for the bunkering stations from the 1860s, and a whole range of finished goods entered the port via Britain and India. Many goods, furthermore, arrived in vessels belonging to the major British shipping lines for whom Singapore provided a convenient station on the routes to the Far East.

That close connection was equally reflected in the early pattern of exports. Thus the region was able to provide a range of products which were important to the British domestic market, ranging from spices to precious metals, as well as important goods which fed into the country trade for use to exchange against products (notably tea) purchased on the China market for use in Britain and Europe. As with imported products, such goods, often products of the local coasting trade, were collected by the major British agency houses in Singapore and loaded onto ships belonging to the British lines. Thus the basic geographic patterns of imports and exports reflected the complex interactions between producers, collecting merchants, agency houses and the shipping lines. The connections between the circuits of capital, investment and trade were mutually reinforced as Singapore grew.

Whilst in the early stages of the development of the port these geographical links were very much with Britain and the British Empire, they were less significant in the later growth of the trading networks. Analysis of the returns suggests that the proportional importance of trade (both imports and exports) with Britain peaked in the late nineteenth century at around 70% of all trade. In later decades the figure rarely rose above 30%, as Figure 7.3 makes clear.

The nature of the goods imported and exported through Singapore is in many respects as significant as the overall volumes and direction of trade. First, the patterns of those goods can be used to examine the extent to which the expansion of the port depended to a greater or lesser extent on just one or two key commodities which were demanded on either regional or global markets. How important, for example, were goods such as tin, rubber, palm oil and timber in the shifting export profiles of the city? Second, information on products can be used to gauge wider changes in the economic structure of the city. Thus the changing balance between primary, unprocessed commodities and processed or manufactured goods may be a useful barometer of the developing economy of the city. A rising proportion of the latter may be a pointer to a more mature and sophisticated economy in the port and city over time.

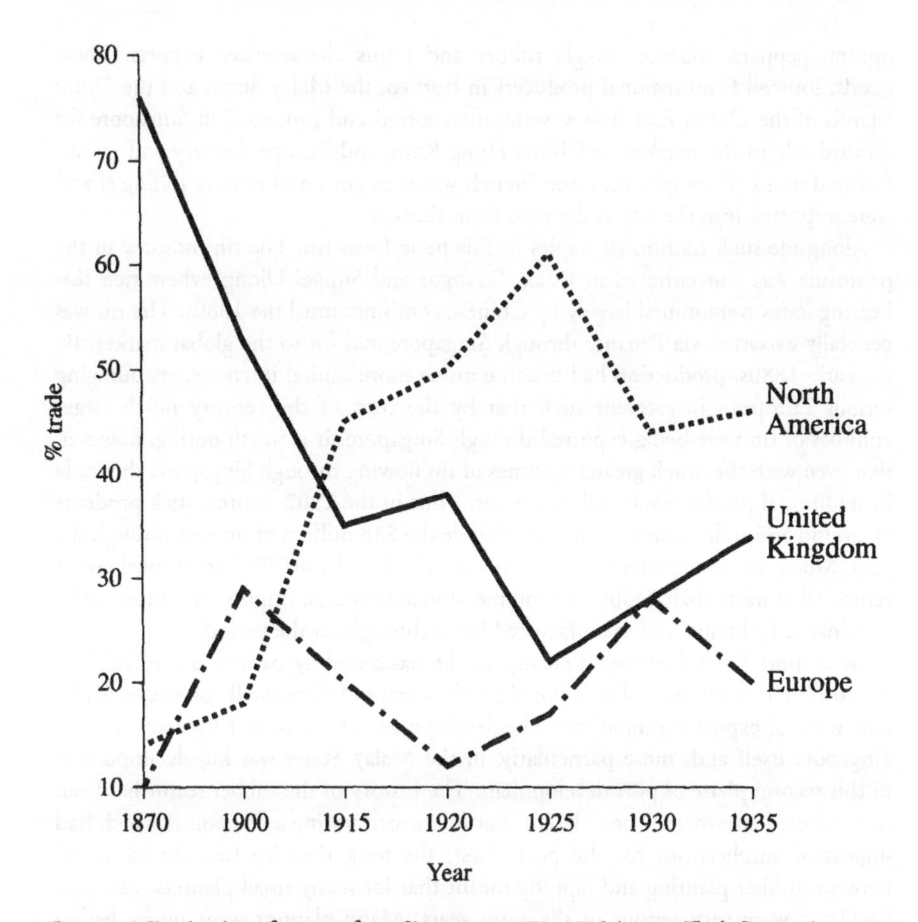

Figure 7.3 Destination of exports from Singapore, 1870–1935 (*source*: Trade Returns).

In examining product profiles from the trade returns, it is evident that the growth of the port reflected its pivotal position as a channel through which key global commodities passed in one form or another. Of special significance were tin and rubber, which dominated port dealings in much of this period. Tin was crucial to a range of industrial processes from tin-plating through to the electrical industry and food processing, whilst plantation (or para-rubber to distinguish it from jungle rubber found naturally) was to be essential to a range of new industries at the turn of the century, most notably the automobile industry.

In examining the products shipped through Singapore, two major phases can be identified, 1870–1910 and 1910–1940. The first phase was marked by an essentially traditional, colonial trading pattern characterised by two features. First, was a heavy reliance on the suite of local products which marked many of the colonial ports of the region such as Penang and Batavia. Local spices and precious metals, dried fish, opium, peppers, tobacco, jungle rubber and resins characterised exports. These goods, sourced from regional producers in Borneo, the Malay States and the Outer Islands of the Dutch East Indies, were often sorted and processed in Singapore for onward sale to the markets of China, Hong Kong and Europe. Luxury and manufactured items (from glassware and French wines to guns and railway rolling stock) were imported into the city and region from Europe.

Alongside such traditional staples in this period was tin. The tin industry in the peninsula was concentrated in Perak, Selangor and Sungei Ujong, where rich tin-bearing lodes were mined largely by Chinese combines until the 1880s. The tin was generally exported via Penang, through Singapore and on to the global market. By the early 1880s, production had become much more capital intensive, encouraging serious European investment such that by the turn of the century much larger volumes of tin were being exported through Singapore. It is worth noting, however, that even with the much greater volumes of tin flowing through Singapore, the trade in traditional products was still dominant. Thus in the 1902 returns, such products at around $86 million were more than double the $38 million of tin sent through the port. Much the same pattern held for most of the decade: in 1907, traditional goods remained at more than double the volume of metal exports. Imports continued to be dominated by luxury and manufactured items throughout the period.

By around 1910, however, a change in the basic trading pattern is evident. The proportional importance of traditional goods began to fall markedly on the back of a new wave of export commodities. The development of the rubber industry in both Singapore itself and, more particularly, in the Malay States was hugely important to this second phase of port development. The history of the rubber sector has been fully covered elsewhere (Voon, 1976), but it is worth noting a few points which had important implications for the port. First, the long time-lag (usually six years) between rubber planting and tapping meant that for many small planters cash-flow problems were very serious in the early years. Many planters went under before reaping the returns of their investment and, from an early stage, large European investors played an important role in buying up and amalgamating plantations. Many of these investors, especially the so-called agency houses who provided technical and financial services to the plantation owners, had major interests in Singapore and the rise of the rubber industry was to bring huge benefits to the city.

As the leading port for the Federated Malay States, Singapore became the trading and financial centre for the rubber industry. Kuala Lumpur was unable to rival Singapore for such services, at least not until the late 1930s. Singapore had by far the best port facilities in the region; Port Klang, the new port for Kuala Lumpur, was not fully established until the 1920s, and Singapore had the right sorting and baling facilities for onward export of rubber to Europe and the United States. By the early 1920s, it had its own exchange for the sale of rubber removing the necessity to act solely through the London Exchange.

The development of rubber production, coupled with the continued role of tin, was the central event in shifting the port from its earlier, colonial, pattern to a much more global role. By 1920, some 38% of all exports by value from Singapore to

Britain were of plantation rubber, whilst the figure for the rapidly-developing US market was even higher at nearly 70%. If the figures for tin are included, the percentages rise to 76% and an astonishing 97%, respectively. Alongside these burgeoning exports, the development of the rubber plantation sector created huge import demands from machinery for the estates themselves to the fine wines and foods with which the management aristocracy entertained themselves at the opulent Raffles Hotel in the city. The rise of rubber, coupled with the staple of tin and, a little later, the oil palm sector, were thus crucial in reorientating the trading volumes, products and direction of the port in at least two ways—in bringing about a decline in traditional jungle products and in encouraging a shift away from the earlier dependence on the British market. With the new focus on the industrial markets of the United States and, from the late 1920s, Japan, the colonial ties were steadily being loosened.

The transformation of the Straits region in this period is thus inevitably dominated by the role of the colonial powers. The extension and consolidation of British and Dutch power in this period was to lead to both the decline of traditional Malay polities on both sides of the Straits and, from 1824, the very division of the Straits themselves into two competing zones. But alongside the more visible, 'European' element in the region lay important networks of trade, culture and political influence which meant that the power of indigenous groups in the region, whether exercised through economic or political currents, remained strong. Those currents, much less visible, nevertheless existed and were to re-emerge with the growth of nationalism and the economic changes of the post-1945 years. It is to those issues that the following chapter turns.

Part 4

COLLECTIVE OPPORTUNITIES

8

ECONOMIC DEVELOPMENT AND POLITICAL CHANGE IN THE CONTEMPORARY PERIOD

The global recession of the 1930s had serious impacts on the economy and trade of the Straits region. As the preceding chapter has emphasised, the emergence of Singapore as both the dominant settlement in the region and as a major global port had been built on the export of key commodities such as tin and rubber, and the import of industrial products into the Malay Peninsula. The trade in both was to suffer as a consequence of recession in Europe and the United States. Indeed, it was not until the late 1930s that the trading volumes through Singapore approached the level of the late 1920s and the decline in the economic opportunities in rubber in particular impacted on all aspects of the region's development.

Government revenues also fell in the 1930s as the value and volume of revenues from export staples fell. This was reflected in a range of impacts. Infrastructural spending, especially on railways and roads, tailed off. This expenditure had been important in linking coast and interior and in facilitating the movement of goods, products and people between the interior and the coast. This was a factor which influenced both the Malay and Sumatran sides of the Straits for the decline in rubber, tobacco and palm oil revenues hit equally hard there. The network of roads and railways which tied the plantations of the Deli region to the Sumatran ports was expensive to maintain and revenue falls impacted sharply on these vital links.

THE JAPANESE INVASION

The Japanese had long had important strategic and commercial interests in the Straits region. The Straits themselves were a vital commercial artery for Japanese shipping and, with the economic resurgence of Japan from the early 1920s, Japanese ships had been taking an increasing proportion of the trade through the major ports of the region, especially Singapore. With important reserves of timber, tin and palm oil, the resources of the region were an important factor influencing Japanese policy, and Japan's wartime aim of establishing a so-called Greater East Asia Coprosperity Zone anticipated the absorption of the reserves, both human and physical, of the region into the Japanese imperial economy. In addition, control of the shipping lanes through both the Malacca and Sunda Straits was vitally important to protect Japan's military and commercial interests.

Whilst Japan's growing economic importance, and potential military threat, was recognised by British military strategists, it is arguable that not enough was done to ensure that Singapore in particular was properly defended. Certainly, British military and naval strategy for the region was predicated on the assumption that Singapore would be the pivot for the defence of British territories east of Suez: as Neidpath (1981, 9) points out, 'the centre of gravity of the British Empire . . . lay East of Suez, especially in the Indian Ocean, which could have been described as a British lake since the three keys to that Ocean—the Cape, Aden and Singapore—were in British hands'. From the early 1920s, a range of proposals and counter proposals for naval installations and military defences at Singapore were discussed (McIntyre, 1979; Neidpath, 1981). But the driving force in London seems to have been the need for defence economies. The naval installations were greatly pared down in the 1920s and early 1930s, as were the artillery and garrison defences of Singapore Island: for McIntyre (1979, 53), 'the Singapore strategy was born out of economy and nurtured in parsimony'. Most analysts anticipated a seaward invasion from the south rather than a landward invasion through the difficult terrain of Malaya. In the event it was the latter that occurred, and the speed with which the peninsula and Singapore itself fell is testimony to the immense surprise at the direction of the Japanese invasion.

The Japanese invasion of Malaya began on 8 December 1941 and such was its success that Singapore fell on 15 February 1942 and, within a month, Sumatra, Java and most of the Dutch East Indies had also capitulated. Singapore became the centre for a new administration which incorporated most of pre-war Malaya together with Sumatra, thereby reuniting for the first time since 1824 states and peoples on both sides of the Straits of Malacca. The successes of the Japanese military machine were outstanding, for nowhere in the region did the invading armies receive significant material aid from pre-war nationalist movements. The 'amazing success of the Japanese invasion,' notes Hall (1987, 860), 'and the rapidity with which it was achieved, did irreparable harm to western prestige'.

Whilst the longer-term aims of the Japanese may have been to create a new, Asian economic bloc in the region, and to rid the region of European and American interests, in the short term little mattered beyond maximising the Japanese war effort. Economic exploitation was harsh and repressive and the economies of both Malaya and Sumatra were ruthlessly plundered. No long-term investment in the tin or rubber industries was undertaken: short-term maximisation predominated. Both rubber estates and the tin industry were soon in a state of disrepair and shipping lines were badly disrupted. The old pre-war lines of the colonial powers—the P and O and KPM, for example—disappeared as their ships were sunk, steamed back to Europe or were requisitioned by the Japanese authorities (Tregonning, 1967). The objectives of the Japanese powers became increasingly short term and pragmatic, objectives which were often accompanied by ruthless security drives, notably against the sizeable Chinese population in the region (especially in Singapore) who were regarded as communist sympathisers.

POST-WAR POLITICAL CHANGE

There is insufficient space here to outline in full the complex and protracted pattern of political change in the Straits region in the post-war period, but without doubt the overwhelming theme is one of independence and the development of nationhood. That development was also to pave the way for successful economic growth in the post-war period.

Within a few months of the Japanese departure from the region in the late summer of 1945, the old division between the countries abutting the states was reimposed. The temporary unity between Sumatra and Malaya was quickly broken as the Dutch sought to re-establish control over the East Indies and the British went back into Malaya and its chief port, Singapore. But it was soon clear that the old pre-war pattern of colonial control was not going to be appropriate for the post-war world. Nationalist groups in both Malaya and the East Indies had learnt much during the Japanese occupation and, with tacit Japanese support, had built up their organisation and experience (Andaya & Andaya, 1988; Osborne, 1988).

In the East Indies, the Dutch attempt to restore its former colonies quickly met with resistance from strong nationalist forces. Groups in Sumatra and Java were especially active in pressing for change and, following a period of internal strife, the Indonesian Republic, first proclaimed in 1945, was established with the support of the international community in 1949. Under the new constitution Sumatra became a constituent member of the republic. Central authority remained in Java, with Jakarta as the capital of a strongly centralised republic. In many respects, then, the position of Sumatra remained similar to that of the pre-war period. Its economy made a major contribution to the national budget, it maintained the impress of colonial economic systems with the huge plantation sector around Deli, and it remained politically subservient to the new Javanese-dominated republic (Tate, 1979). Tensions between core and periphery—whether between Sumatra and Java, or between regions such as the territories of Aceh and Minang—and the centre remained unresolved in the political solutions of the post-war republic.

In Malaya there were a complex series of proposals and counter proposals in the immediate post-war period to provide a new constitution and shape to Malaya as a preliminary to independence (Andaya & Andaya, 1988; Tarling, 1993). Problems over citizenship rights for the different ethnic groups, over the powers and precedence of the sultans, and over chronic communist insurgency in the late 1940s and early 1950s, created serious political and social problems. A new constitution and formal independence were not ratified until August 1957, with Singapore and the west coast states along the Straits assuming the role of constituent states in the new federation. The Borneo states became part of a new Federation of Malaysia in 1963, but the position of Singapore remained problematic. The tensions between Singapore and the new federation were both political and ethnic. Singapore was, of course, predominantly Chinese and arguably more 'left-leaning' than much of the rest of Malaya. Together with that other Straits Settlement, Penang, it had experienced long years of in-migration from southern China and had a very different social and ethnic composition. Its role as a port also gave it a rather different economic and social dynamic from the rest of the federation and tensions ultimately led to Singapore seceding from Malaysia in 1965 to become an independent nation state (Turnbull, 1989).

The political settlements in the states bordering the Straits have been accompanied by a range of social and political issues influencing their social and economic development. It is hardly suprising that the ethnic diversity of the region—a product of its long openness to in-migration over the centuries and the diverse influences—Indian, Chinese, European, which have flowed into the Malay world—has brought with it both strengths and weaknesses. The intrinsic ethnic diversity of Malaysia with its Malay, Chinese and Indian population groups has shaped both political and economic strategy. After ethnic riots in Kuala Lumpur in 1969, the New Economic Policy (1971–1991) was devised to try and achieve greater harmony between ethnicity and wealth, with a range of consequences for different population groups (Cleary & Shaw, 1994; Idriss, 1990; Jesusadon, 1990). The New Development Plan which followed the NEP had a similar mix of growth-orientated and redistributive policies for the diverse ethnic groups of the nation.

Ethnic diversity has also underpinned other political and economic issues in the region. The absorption of Sumatra into the Republic of Indonesia, whilst in many ways continuing the pattern of governance instituted under the Dutch, did place many independently-minded regions under the direct control of Jakarta. Thus, for example, the old Sultanate of Aceh, one of the most powerful in the region between the sixteenth and eighteenth centuries, now the Aceh sub-region, and parts of the Minang region in Sumatera Selatan, have been subject to unrest over their relations with Jakarta. The Aceh region, in particular, has been under internal security orders for lengthy periods of time since the establishment of the republic.

Similar centre–periphery tensions characterised Malaysia. The secession of Singapore from Malaysia in 1965 reflected both ethnic differences and tensions between a region wedded to global trade and the outside market and (at the time) a more rural, inward-looking hinterland. The contrasts between the coastal and inland polities has, of course, been an important recurring theme in the development of the Straits region. The state of Penang has also exhibited some differences (and tensions) with the rest of the country, a reflection of its greater proportion of Chinese and, perhaps, a historic legacy of openness to the cultural and economic currents flowing through the Straits.

PATTERNS OF ECONOMIC DEVELOPMENT

Despite the rapid reversion to a political division between the Straits regions, common interests, and to some extent common destinies, continued to unite the different political units. First, there was the evident problem of wartime damage. There was widespread physical damage throughout the region. Port infrastructures, especially those of Singapore, had been damaged both through the physical impact of bombing and warfare, and through the lack of any long-term investment. In Singapore, for example, there had been relatively little investment anyway during the recession of the 1930s and this lack was further accentuated by the impact of the war. Elsewhere, many of the other ports—Penang, Balawan, Malacca, Port Dickson—were similarly damaged, with ruined infrastructure and shipping lanes blocked with damaged ships.

The nature of the Japanese occupation was highly exploitative and many of the plantation estates were in a ruinous condition, contributing to the difficult economic conditions. Emergency food production had been a key priority in both the Malay States and in Sumatra, and intensive production of food crops took priority over the long-term maintenance of plantations. Rotation systems were abandoned, there was little maintenance of plant and machinery, and a number of plantations in the Medan district were broken up into small food producing plots for landless labourers (Pelzer, 1978, 122–125). In the struggle for Indonesian independence, military groups were able to assume control of many of the Medan plantations using cash from sales of plantation land to fund the military effort.

Infrastructural damage was also considerable. Allied bombing and military advances had caused major damage to the urban fabric of towns and cities, especially in Malaya, and road and railway connections had been disrupted. All these factors created major difficulties for the restoration of political and economic order; add to this the military struggle between the Dutch and Indonesian forces in Sumatra, and the emerging political conflicts over independence for Malaya, and it is clear that the immediate post-war years were very difficult for the region as a whole.

It was not just the extent of wartime damage that the region shared: there were a number of other features of the economy which were to remain as legacies of the colonial period long after the region had achieved independence. The extent to which the region exhibited the classic 'dual economy' pattern postulated by Furnivall (1939) has been a matter of much discussion. It is arguable that the economies of both coastal Malaya and Sumatra had been strongly shaped by external demands for key products, and that this had created a relatively dynamic export-crop economy alongside a moribund indigenous agricultural sector. Reliance on the import of the industrial products needed to sustain the export economy, coupled with the import of almost all consumer products, had also, it is argued, stunted the emergence of a viable industrial sector in these economies.

There is some evidence to sustain this view of the economies of the region in the post-war period. As Chapter 7 has suggested, one of the features that distinguished the economic development of the European colonial period from earlier phases in the region's history was the extent to which the production process of the new commodities demanded by the global economy was shifted from indigenous to external control. The production of rubber, tea, coffee, oil palm and tin, products which remained essential to the economic survival of the region well into the late twentieth century, were for a long period of time in foreign hands. Reliance on external investment, strong external control of the production of export staples and the repatriation of profits remained economic features of the region until relatively recently.

In the agricultural sector one of the key problems facing the administrations in both Sumatra and coastal Malaya was the nature and control of the vital plantation sector. Independence was to pose particular questions for the plantation sector in Sumatra. Strong anti-Dutch sentiments culminated in the nationalisation of all Dutch estates in 1957 and they were placed under government control along with other foreign-owned estates in 1962. But the political changes after 1966 brought more favourable conditions for foreign investors, and by the 1980s foreign ownership accounted for around one-quarter of all estates in North Sumatra, with the government remaining in control of the majority (Cleary and Eaton, 1996; Courtenay, 1980).

In Malaya the plantation sector had been a major target for communist insurgents during the 1950s, but foreign ownership remained very important in order to ensure high levels of investment in a sector which, as in Sumatra, had experienced chronic under-investment over the past decade. Independence in 1957 did not bring any major changes, although there were increased sales of plantations to local interests. In 1963, around 60% of rubber plantations were foreign owned; by 1973, the figure had fallen to around 40% and has changed little since then (Courtenay, 1980, 200). Federal and state development corporations, notably the FELDA (Federal Land Development Agency) have played an increasingly important role in developing the plantation sector, especially in oil palm, with large investment in plantations in the Johor region which have transformed the physical appearance of large swathes of land abutting the main highway from Singapore.

Alongside the reconstruction of the agricultural sector—vital in the drive to develop export revenues and foreign earnings—administrations faced difficult problems in regenerating the important mineral exports. The tin industry in Malaya, an important component of the economy in the pre-war period, was in a poor state and required major external investment to improve plant. It was not until the late 1940s that production began to recover (Tregonning, n.d.). The development of an indigenous industrial base was a much more difficult, long-term problem. Through the 1920s and 1930s, vast quantities of industrial goods had poured through the port of Singapore to meet the demand of both the city and the Malayan plantation sector, as well as burgeoning consumer demand. But the bulk of that demand was to be met through imports rather than the development of indigenous industry. Whilst Singapore and the growing federal capital, Kuala Lumpur, had begun to develop local industry, particularly in textiles and food processing, large-scale manufacturing industry was absent and it was raw materials that were exported, rather than processed goods. This pattern tends to underline the dual nature of the colonial economy in the immediate post-war period in the Straits region.

EARLY INDUSTRIAL GROWTH

In both Peninsular Malaya and coastal Sumatra, small-scale industrial and artisanal activity had long been an important part of the domestic economy but, as we noted earlier, most economic activity was focused on the production of primary exports which were reprocessed elsewhere. Indeed, the economic and political structures of the region mitigated against the development of large-scale manufacturing industry. Industrial development in the two decades following the end of the war had tended to follow pre-war patterns. The main focus of development was twofold. First, there were major efforts made to recover production in the key pre-war export sectors—rubber, tobacco, tea, coffee, timber—as well as efforts to expand products such as oil palm, minerals and hydrocarbons, which were seen to hold major potential for export growth. The recovery of exports was seen as crucial to increase the revenues needed for the urgent infrastructural and social development required in the region. In Malaya, the expansion of the coastal plantation sector, coupled with infrastructural upgrades, paved the way for the emergence of important growth areas in Singapore itself, Johor Bahru and the Klang valley between Kuala Lumpur and the rapidly developing Port Kelang. In Sumatra, nationalisation of the plantation sector coupled with the break-up and redistribution of some estate lands led to a very sluggish recovery to pre-war export levels. In addition, the climate of political uncertainty led to some flight of foreign capital from Sumatra.

A second feature of this period was the start of attempts to develop indigenous industry. The establishment of an independent Malaysia in 1957 galvanised efforts to reshape industrial and export policy in order to reduce the strong dualism in the economy. One of the main foci of policy was the establishment of import-substitution-industrialisation (ISI) as a key strategy (Idriss, 1990). Thus, alongside efforts to reshape and diversify the agricultural sector, attempts were made to establish an industrial sector producing goods which had formerly been imported. Products such as textiles, processed food and drinks and simpler manufactured goods were seen as important in replacing some imported goods and thus aiding the overall balance of payments. Singapore, rapidly emerging as an important and dynamic manufacturing centre, began to develop such industries in the late 1950s. A large, low-cost labour force, together with a sizeable consumer market were key advantages (Rodan, 1987). Elsewhere in the Straits region, Penang was also beginning to establish a viable industrial base in this period and the lower Klang valley was emerging as an industrial centre to serve the rapidly growing federal capital. The development of the new town of Petaling Jaya from the mid-1960s aided this growth.

On the other side of the Straits, industrial growth in Aceh, Jambi and Riau was sluggish for both political and economic reasons. With the heavy concentration of both political and economic power in Java, the main role of Sumatra continued to be the production of the primary unprocessed goods which provided the export revenues for investments generally made elsewhere. Thus in the post-war period a clear divergence in the economic destinies of the regions on each side of the Straits began to etch itself on the economic landscape: a number of dynamic centres on the east from Georgetown, Penang in the north, through the Klang valley to Singapore; whilst to the west an economy heavily dependent on primary goods and agriculture, lacking infrastructure and facing both economic and political tensions with the centre.

Thus in 1960, for example, the Singapore economy had 26% of its working population employed in the industrial sector, compared to figures of around 6% for Malaysia as a whole and almost negligible proportions for Sumatra. Even that pattern was geographically skewed: as Leinbach (1972) showed in mapping what he termed 'modernisation surfaces' in Peninsular Malaysia, the pattern of development (even allowing for the inadequacies of the data) was highly skewed to the port city of Penang, Kuala Lumpur and the Klang valley, the tin mining areas of Ipoh and Singapore. Skewed, then, towards the coast and the ports: a pattern reflecting the continued importance of the Straits in channelling economic and social change.

THE GROWTH OF MANUFACTURING

By the early 1960s, the economies of both west coast Malaysia and Singapore, whilst having developed some significant industrial growth, were still heavily dependent on essentially colonial patterns of growth, servicing the core global economies through supplying key raw materials. For northern Sumatra much the same patterns applied, with the region filling the role of a satellite economy for Java/Jakarta through the supply of plantation products and hydrocarbons. As noted above, both Malaysia and, after independence in 1965, Singapore, had actively pursued ISI policies, stimulated in Malaysia by a Pioneer Industries Policy which provided government help for new industries. By the late 1960s, Singapore and, a little later, Malaysia embarked on aggressive export-orientated-industrialisation (EOI). This was based on establishing competitive manufacturing industries, usually through attracting multinational capital, and taking advantage of low wage rates in the region (for a good, general account of the process, see Dicken, 1992; Rigg, 1997).

For Singapore, this shift in strategy marked the beginning of a long period of sustained industrial growth which was to transform the economy into a powerful Asian 'tiger economy' by the mid-1980s. A strong stable government under the inspired leadership of Lee Kuan Yew, combined with careful labour and social legislation, transformed the economy of the state developing an outward-looking global marketing strategy. Initial growth took advantage of low wage rates to develop textiles and manufacturing; from the mid-1980s, a focus on high value-added products in electronics, biotechnology and computing software has continued to transform the economy. With a highly educated workforce and close cooperation between the state and organised labour, continuous innovation and policy flexibility has ensured high growth rates throughout the period, even with the currency and economic difficulties of the mid-1990s (Rigg, 1991; Rodan, 1987). Certainly by the mid-1980s, Singapore had emerged as the clear economic powerhouse of the region, with a position of primacy based not only on its continued strategic and commercial importance as a port, but on its position as a major, innovative producer. In addition, the large capital reserves available to the state through a range of compulsory retirement funding schemes (notably the Provident Fund) meant that the state had begun to play an increasingly important role as a source of foreign investment capital and joint ventures for the region as a whole. Its growth rates rarely faltered during this

Table 8.1 Economic growth rates in the Straits region

	GDP average annual growth			*GNP/capita (US$)*	
Country	*1970–1980*	*1980–1990*	*1990–1994*	*1997*	*1994*
Indonesia	7.2	6.1	7.6	4.7	740
Malaysia	7.9	5.2	8.4	7.8	3,140
Singapore	8.3	6.4	8.3	7.8	19,850

Source: Institute of Southeast Asian Studies (1999). *Regional outlook 1999–2000*. Singapore: ISAES. Rigg (1997, 6).

period and, even during the economic crisis of the mid-1990s, the state maintained good levels of growth.

Alongside the expansion and transformation of port facilities in Singapore, dealt with in a later section, the city and island has been the subject of enormous replanning and reconstruction. The Singapore Urban Redevelopment Authority (URA) has been the main single agency responsible for urban transformation. At the heart of urban development were a series of public housing projects which, from the late 1960s onwards, have placed large numbers of Singaporeans in a range of housing projects. In addition, huge investment in infrastructure and transport facilities, including the development of a Mass Rapid Transit System, the expansion of public transport, and the establishment of a number of new towns on the fringe of the urban area has created an urban fabric almost unrecognisable from that which existed at the end of the colonial period (Yeoh, 1996; Yeoh & Kong, 1995) (Figure 8.1). Today the 3.0 million population of the city state live in some of the best urban conditions in the world.

Malaysia was to pursue similar strategies to Singapore: for Lim (1985, 37) the late 1960s saw 'the transformation of the state from a facilitator of capital accumulation to a direct actor in that process'. The development of a strong export-orientated manufacturing sector, based initially on low wage costs, was an important element in reducing Malaysia's dependence on traditional primary exports. In addition, it was anticipated that high economic growth would allow for some redistribution of wealth between competing ethnic groups in the country, a key plank in the NEP programme. By the mid-1970s, Malaysian growth rates were consistently high, built on the basis of both continuing expansion of primary exports (palm oil, rubber, timber) and on the growth of the manufacturing sector. Strong, centralised planning coupled with the maintenance of fiscal incentives for foreign investment made the country an attractive location for Japanese, US and South Korean multinationals. Despite the economic problems of the mid-1990s, overall growth rates still remain high in global terms.

In geographical terms one of the main outcomes of the 20-year NEP was a sustained shift of population from rural to urban areas and, more specifically, the development of key urban-industrial nodes on the west coast of Malaysia. From Penang in the north to Johor in the south, economic growth has created important

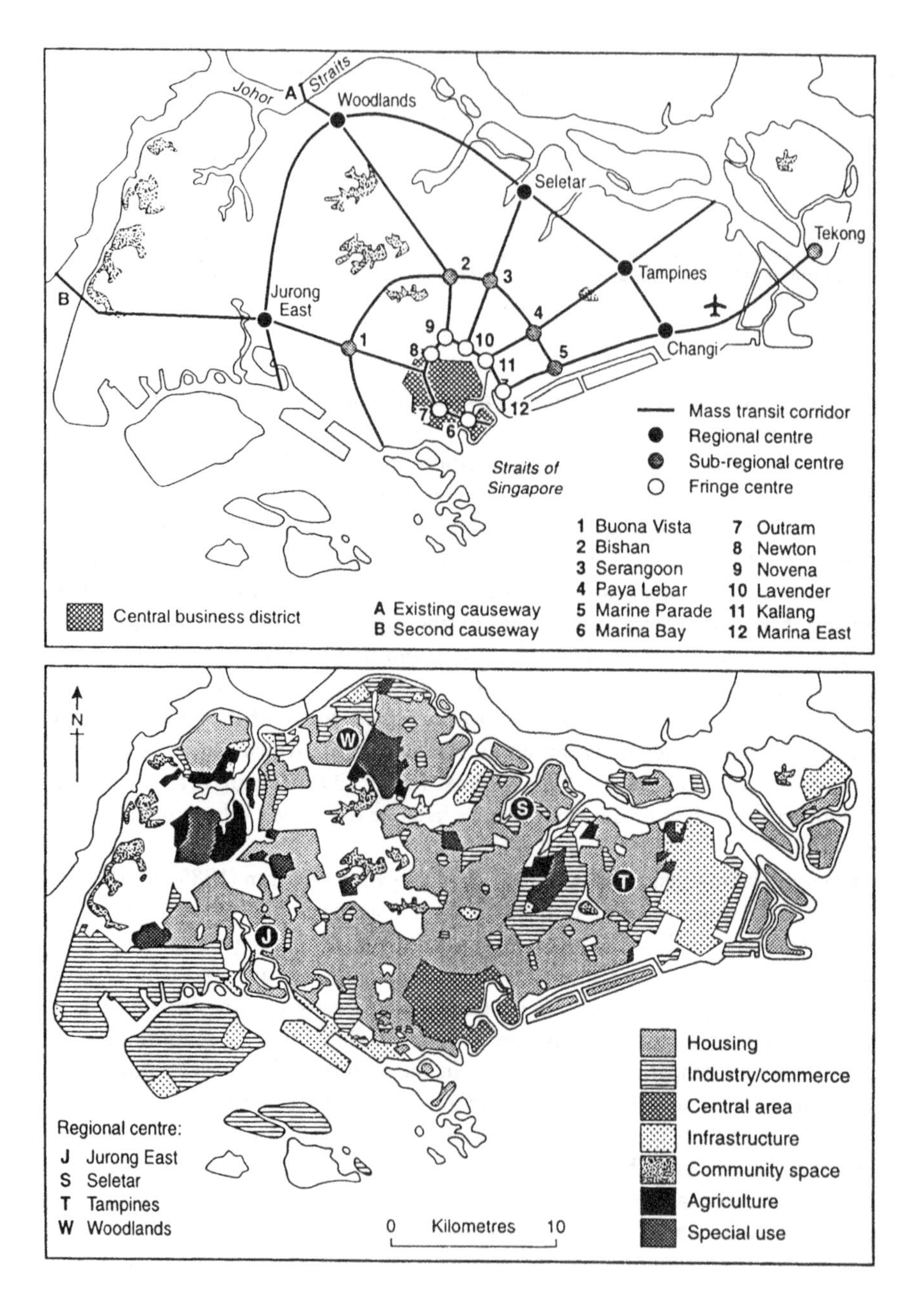

Figure 8.1 Urban structure of Singapore, mid-1990s (*source*: Urban Redevelopment Authority).

Table 8.2 The 'western corridor' in the Malaysian economy

	Western corridor	*% of Malaysia*
Land Area	75,504	49.5%
Population (1990)	11,655	64.7%
Per capita GDP (1980)	3,519	3,343
Per capita GDP (1990)	4,651	4,392
Per capita GDP (2000)	7,562	6,874
Industrial estates (1992)	112	165

Note: 'Western corridor' = Perlis, Kedah, Penang, Perak, Selangor, Negeri Sembilan, Malacca and Johor.
Source: Naidu (1997, 46–47).

development zones. The most striking, the Klang valley, is a corridor of growth linking Port Klang, the most important port in the country, with Kuala Lumpur through the new town of Petaling Jaya. The dominance of this region, together with the industrial centres of Georgetown and Ipoh, over the eastern coast of the country reflects the long historical importance of the Straits of Malacca in shaping development in the peninsular. In terms of per capita GDP, it has been the western seaboard which has dominated (see Table 8.2).

The growth of Penang illustrates the nature of economic and social change on the eastern side of the Straits well. As a Straits Settlement, it had remained somewhat in the shadow of Singapore and, on independence, the high growth rates in Kuala Lumpur had made the island something of an economic backwater in the 1960s. The removal of its free port status in 1968 further accentuated its economic difficulties. The elaboration of the NEP, coupled with the switch to EOI policies were, however, to transform the economy of Penang. The Penang Development Corporation was established to foster economic growth and with an international airport, excellent port facilities and a well educated workforce, the island was well placed to benefit from the new economic conditions. The creation of a Free Trade Zone also facilitated foreign investment and by the early 1980s the Penang Development Corporation had attracted some 250 factories employing close to 60,000 people in the state (Cleary & Shaw, 1994). Despite fluctuations due to changing economic circumstances, by the early 1990s numbers and factories had almost doubled. Textiles and electronics dominate, with most investment attracted by a cheap and docile, largely female, labour force. In the longer term, however, a dependence on cheap labour may create problems, especially as wage rates elsewhere in the region are lower.

Johor, at the southern tip of the peninsula, exhibits similar patterns of growth. Historically, an important sultanate, as we showed in Chapter 7, the economic expansion of Singapore since the early 1960s has had enormous implications for both the state and city. As wage rates in Singapore have risen, Johor has been able to attract industry across the causeway to locate in the state, bringing important economic benefits to the region. A buoyant economy has also been boosted by the large numbers of people, both Malaysians and Singaporeans, who live in Johor and commute daily to Singapore, taking advantage of lower living costs (especially for housing) on the Malaysian side of the causeway.

Across the Straits, the pattern of recent economic growth in east coast Sumatra has been rather different. Of course, the problems of development in Indonesia as a whole are radically different from those of Malaysia and Singapore, with very different endowments of physical and human resources, and very different issues to tackle after independence. It can be argued that in the provinces of Riau and Jambi, North Sumatra and Aceh, historically the regions closest to their neighbours across the Straits, economic conditions remain quasi-colonial in that, the main role of the region continues to be as a supplier of raw materials to fuel more sustainable economic growth elsewhere in the country.

As a supplier of hydrocarbons the Riau province is especially important. With a number of refineries and a major hydrocracker plant, Riau provides over one-half of Indonesia's petroleum. With the main producing area around Duri, much of the oil is exported through the deep water port of Dumai. Off-shore and on-shore fields have been developed and whilst the role of mining is important in terms of GDP, there are relatively few linkages between the oil sector and manufacturing generally (see Figure 4.1). As Rice (1989, 127–128) notes, 'owing to the weak linkages between the gigantic petroleum production and refining sector and other sectors, the economy of Riau could be characterized as dualistic'. Batam Island, to the south, administratively part of Riau though only 20 km southeast of Singapore, remains the exception to this pattern and provides an example of international economic cooperation and development through the growth triangle mechanism examined later.

Aceh to the north, historically one of the most important sultanates in the region, was never really quelled by either the Dutch or, indeed, the Indonesian government. Again, like parts of Riau and Jambi, the economy has been heavily dependent on primary exports of, in this case, liquified natural gas (LNG). The extensive reserves around Lhok Seumawe and Lhok Sukon have been exploited heavily over the last two decades to provide major export revenues for the country. With substantial foreign investment and the establishment of an ASEAN Aceh Fertilizer plant, GDP growth during the 1980s was high. But there is a strong enclave pattern of development around the hydrocarbon region (Hill, 1989, 116), with heavy dependence on external investment and markets. Like Jambi and neighbouring North Sumatra, the prospects for a large, sustainable manufacturing sector as has developed in west coast Malaysia and Singapore appear very remote. The post-war pattern of economic divergence between the east and west of the Straits remains significant.

PORT AND SHIPPING NETWORKS

The Straits remain, as in the past, one of the most important shipping lanes in the world. It has been estimated that they are the second busiest strait in the world after the Straits of Dover. The growth in demand for hydrocarbons in East Asia, coupled with the economic growth of Japan and the 'tiger' economies, have led to burgeoning

Table 8.3 Shipping flows through the Straits of Malacca, 1982–1997

Year	*Arrivals at Singapore*	*Estimated through traffic*
1982	36,361	43,633
1985	36,531	43,837
1988	44,855	53,826
1990	60,347	72,416
1992	81,340	96,000
1993	92,655	110,500
1996	117,723	139,000
1997	130,333	n.d.

Source: Naidu (1997, 34); Maritime Port Authority of Singapore (*Annual Reports*).

use of the Straits by ships servicing those growth markets. Couple this with the economic growth of Singapore and Malaysia, and the continued importance of the Straits in regional and global shipping patterns is clear.

The Straits route represents the shortest route for oil tanker trade between the Persian Gulf and East Asia and the traffic in tankers now constitutes a major part of both through and transit traffic. Currently, a large tanker is likely to take an average of 36 hours to transit the Straits from the Andaman Sea to the South China Sea and, with a lowish estimate of around 200 vessels using the Straits each day, a tanker would typically pass around 150 vessels moving in the opposite direction, an average of one every 14 minutes. Around 72% of all east-bound laden tankers use the Straits of Malacca route compared with 28% shipping through the alternative Macassar and Lombok Straits. This alternative route is plied mainly by supertankers unable to comply with the maximum 3.5 m underkeel clearance limit currently operating in the Straits of Malacca. This alternative route is longer and more expensive. One estimate suggests that using the Lombok–Macassar Strait adds around 1,000 nautical miles to the typical Gulf–Japan route, an added journey time of some three days. Only for the very largest supertankers in excess of 300,000 dwt would this be economically viable. If, for strategic or navigational reasons, all tankers were required to follow this route the estimated annual cost would be between US$84 billion and US$250 billion (Hamzah, 1997, 114).

It is difficult to estimate traffic volumes in the Straits because of the huge variety of ships using the routeway. In addition, by no means all ships transit at the major ports of the Straits, especially Singapore. Naidu (1997, 33–34) has estimated that, in the mid-1990s, around 300 ships used the Straits daily; but that figure would probably double when account is taken of the variety of smaller vessels—from fishing boats to pleasure craft—that criss-cross the waters. As Naidu has also shown, the numbers of vessels using the Straits has continued to grow, almost doubling between 1982 and 1991.

The Straits, then, not only serve as an important routeway for the tanker trade, they also constitute a major convergence point for the important east- and west-bound shipping services that service the economic blocs of Europe, North America

Table 8.4 Major shipping incidents in the Straits, 1977–1994

	Number of ships	%
Type of casualty		
Collision	25	35
Grounding	13	18
Explosion/fire	5	7
Foundering	7	10
Others	21	30
Ships involved		
Container carrier		4
Cargo carrier		32
Passenger liner		8
Oil tanker		17
Fishing vessel		15
Tug boat		7
Others		17

Source: Port of Singapore Authority. (*Annual Reports*).

and East Asia. Singapore is the key articulation point for this trade: as Robinson (1997, 265) notes, it 'sustains two major shipping networks—one to the east and into the South China Sea and the Pacific, and the other to the west and the Indian Ocean and via the Suez, to Europe'. Around 80% of all vessels using the Straits call at Singapore.

Increased demands for oil, coupled with strong regional economic growth, have meant expanded use of the Straits by both regional and global shipping lines. This rapid growth in the use of the Straits has led to a range of navigational and transit issues arising. Safety of navigation is probably the key issue, especially in the light of the dangers of oil spillages from tanker incidents. There were an estimated 71 major 'shipping casualties' between 1971 and 1993 on the Malaysian side of the Straits: almost one-fifth of these involved oil tankers. Three major tanker accidents alone—the *Showa Maru* (1975), *Diego Silang* (1976) and *Nagasaki Spirit* (1992)—resulted in the spillage of more than 30,000 tons of crude oil, with huge impacts on marine ecosystems. Table 8.4 indicates the nature of accident incidents in the Straits in the recent past. Chapter 10 addresses some of these difficult environmental issues, but the problem of traffic control and regulation is a difficult and chronic issue.

A Traffic Separation Scheme (TSS) operates in the Straits off One Fathom Bank in the north and in the northwest approaches to the Singapore Strait in the south (Figure 8.2). In the Singapore Strait itself, traffic separation schemes also operate to guide shipping through the complex network of islands separating Singapore Island itself from Pulau Balam and Pulau Bintam. The establishment of the TSS required complex negotiations between Indonesia, Malaysia and Singapore, together with the International Maritime Organisation. Elsewhere in the Straits, however, ships are free to plot their own course. With strong winds, shallow channels, numerous unmarked wrecks and tight tidal conditions for deep-draught vessels, the probability of accidents

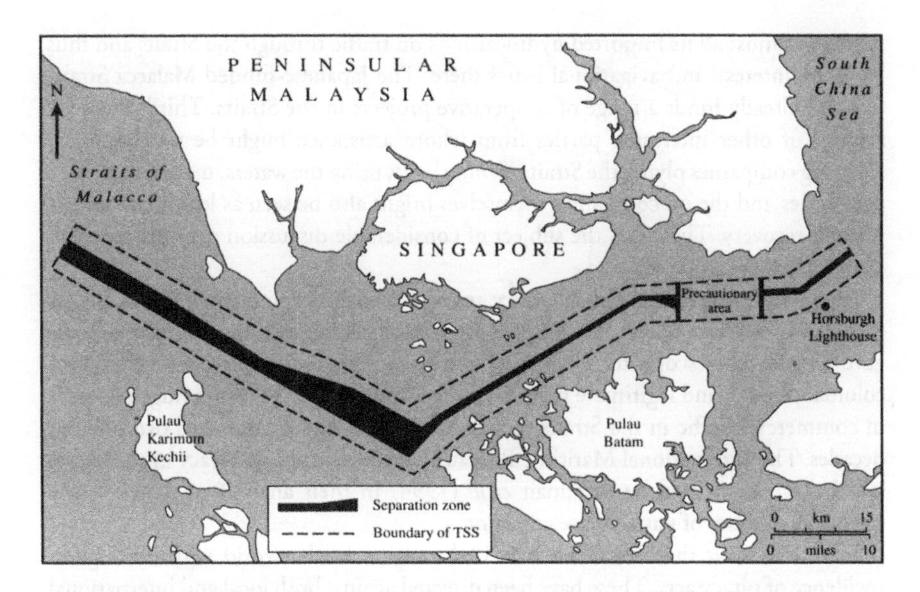

Figure 8.2 Traffic Separation Scheme in the Straits of Malacca and Straits of Singapore.

occurring is high and this has led to demands for greater regulation and control of shipping routes in the Straits.

The negotiations for the creation of a TSS were complex in that they impinged on problems of legal definition: were the Straits to be regarded as 'international' and hence subject to the same regime as the high seas or were they, as argued by Malaysia and Indonesia, territorial waters within the definitions put forward by the United Nations. Arising from these legal debates are issues connected with the control of shipping through separation, the possibility of tolls, and the extent to which shipping passing through the Straits should contribute to the costs of upkeep of navigational and other aids. Both Malaysia and Indonesia have proposed various extensions to the TSS, as well as the possibility of establishing a special in-shore traffic zone for local shipping along the Malaysian coast. Equally important have been proposals to make better navigational beacons available, together with greater use of radar to monitor and control shipping in the Straits.

One of the central problems confronting navigation improvements in the Straits is the question of cost. Quite simply, who should pay? As Peet (1997, 153–154) has argued, three different types of interest can be identified. First, are the port states themselves—Malaysia, Indonesia and Singapore. Their interests are not, however, the same. Singapore, as the most important destination port, has perhaps the greatest to lose from ongoing problems in the Straits, whilst the interests of Indonesia may be seen as primarily focused on the important local traffic across and along the Straits. Indonesia, furthermore, might benefit from a diversion of traffic from the Malacca to the Sunda Straits. Second, are the interests of those states that depend heavily on shipping in the Straits for the import of goods and raw materials. Japan, in particular, relies for almost all its imported hydrocarbons on traffic through the Straits and thus has huge interests in navigational issues there. The Japanese-funded Malacca Straits Council already funds a range of cooperative projects in the Straits. Third, there are a range of other interested parties from whom assistance might be sought. Local shipping companies plying the Straits, fishing boats using the waters, marine insurance companies and the oil companies themselves might also be seen as legitimate targets for cost recovery. The issue, the subject of considerable discussion over the past two decades, has yet to be resolved.

Alongside navigational issues, piracy remains a chronic problem. Historically, as earlier chapters have noted, the complex patterns of islands and inlets along the Straits gave considerable scope to a range of activities regarded as piracy by European colonial powers, and legitimate trade by indigenous groups. With the huge increases in commercial traffic in the Straits the problem of piracy has increased over the last decades. The International Maritime Bureau collates statistics on piracy in the region which have been used by Beckman *et al* (1994) in their analysis of piracy in the region. A number of patterns are apparent.

First, it is clear that Southeast Asia is the region in the world with the highest incidence of piracy acts. These have been directed against both local and international shipping. All kinds of vessels—container ships, bulk carriers, tankers and ordinary cargo ships—have been subject to attack: in 1992, for example, some 73 attacks were reported in the region as a whole, with around 50 of those attacks in the Straits themselves or at the entry and exit points (Beckman *et al*, 1994, 31). The data itself is not always reliable and comes from a range of sources, but the report identifies a number of significant common features. Three areas are especially prone to attacks: the area around Pulau Bintan, close to the Phillip Channel at the southern end of the Straits, and in the Straits themselves. The great majority of attacks occur during the hours of darkness and usually result in the ransacking of crew's quarters and the theft of money and valuables. The crew and master are often tied up, with consequent drifting of the vessel. The consequences to other shipping of these attacks are potentially disastrous.

Attempts to police the Straits in order to reduce the dangers of piracy are fraught with difficulties. Shipping traffic is very dense and the maze of islands and inlets, especially at the southern end of the Straits, means that pirate vessels can quickly disappear. In addition, as Beckman *et al* (1994, 14) note, 'many of these attacks are located in areas close to the existing international boundaries or where maritime boundaries have still to be delimited and agreed upon,' making the task of policing yet more difficult. Certainly the effectiveness of the Singapore Police Coast Guard, together with naval patrols, has begun to reduce the problem and there is evidence that overall piracy rates are now falling. Singapore and Indonesia agreed in 1992 to establish direct communications between their navies and to coordinate naval patrols in the Singapore Straits and Phillip Channel. Malaysia and Indonesia have also stepped up their cooperation, establishing a joint Maritime Operation Planning team to coordinate anti-piracy patrols in the Straits. But the issue of maritime borders remains a contentious one and makes 'hot pursuit' by one navy into another country's territory most unlikely.

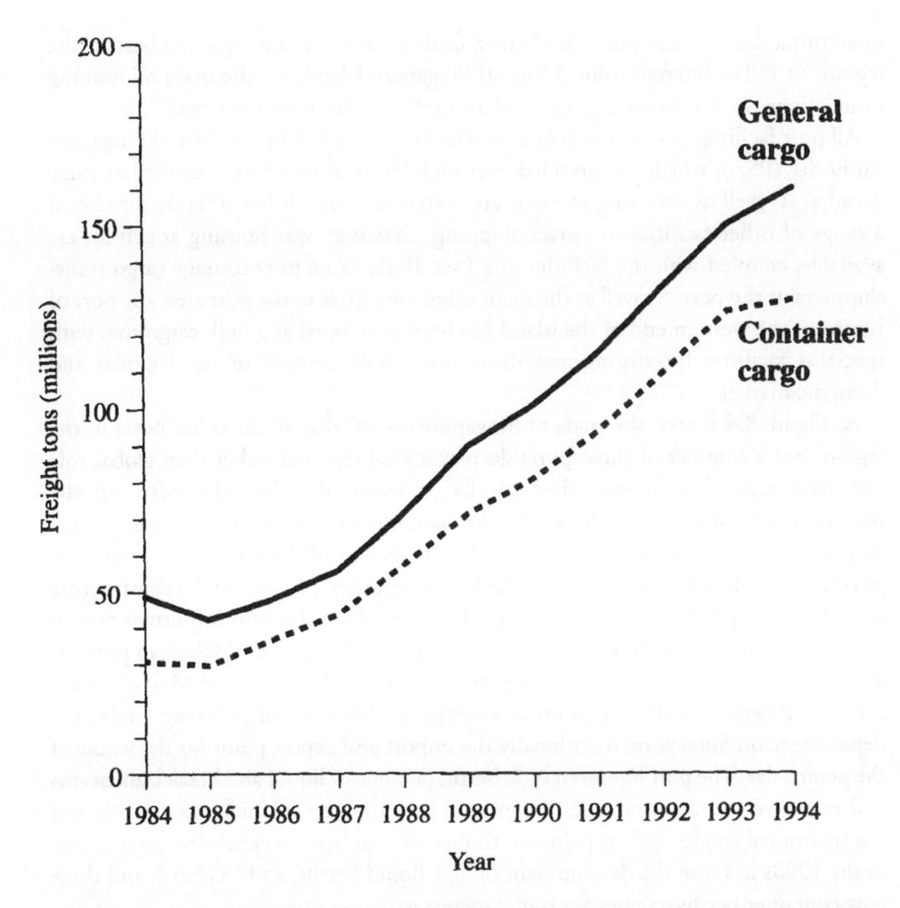

Figure 8.3 Cargo throughput in Singapore, 1984–1994 (*source*: Port of Singapore Authority).

Singapore, the most important port in the region in the pre-war period, remains pre-eminent in the Straits. Globally, it is ranked first in terms of shipping tonnage passing through the port and second in terms of both cargo and container throughput. Over 500 shipping lines and 130 container lines use the port in the course of a normal year (Port of Singapore Authority, 1994). Chapter 7 outlined the emergence of Singapore and its rise to regional and global pre-eminence from its foundation in 1819. The economic growth in the region has been reflected in the continued growth of traffic at the port. As Figure 8.3 shows, there have been major increases in all categories of throughput—vessels, cargo and containers. The expansion of container traffic has been especially striking. The three major container facilities—Tanjong Pagar (seven main berths), Keppel (six main berths) and Brani (six main berths) have been expanded and modernised to cope with traffic increases. An international passenger terminal catering to cruise ships, together with general port cargo facilities for non-container traffic at the Pasir Panjang Terminal, complement the extensive, modern facilities at the port. Bunkering facilities are also amongst the best in the region. At Pulau Bukom, some 5 km off Singapore Island, are the main oil refining installations, with a refining capacity of around 25 million tons per year.

All port facilities and investment strategies are controlled by the Port of Singapore Authority (PSA), which has presided over high levels of investment in the last three decades. As well as investing in plant and cargo facilities, it has recently developed a range of other facilities to attract shipping. Extensive warehousing amenities are available, coupled with the facilities of a Free Trade Zone to encourage cargo transshipment at the port. As well as the main cargo amenities in the port area, the port of Jurong at the western end of the island has been developed as a bulk cargo site, with specialist facilities for storing and distributing bulk cement to the regional and domestic market.

As Figure 8.4 shows, the trade of Singapore dwarfs that of the other ports in the region. But a number of those ports do play a vital regional rather than global role and have expanded facilities dramatically in recent decades. The ports on the Malaysian side of the Straits have, like the economies of the respective regions, outstripped Sumatran ports. That is primarily a reflection of the more rapid economic growth rates, but it also reflects a deliberate strategy of port and infrastructure development by the Malaysian government. Port Klang is the most important port in Malaysia, with more than double the cargo volumes of any other Malaysian port. Its growth has reflected the major changes in the industrial structure of Malaysia since the mid-1960s, as well as a political strategy by Malaysia of reducing Malaysian dependence on Singapore, traditionally the import and export point for the whole of the peninsular. The port has seven bulk berths, a range of liquid and break bulk berths and extensive container facilities. Its exports include both manufactured goods and the traditional staples such as palm oil, timber and rubber. An expansion programme in the 1990s has seen the development of new liquid berths, an LPG berth and three new container berths to cater for traffic increases.

Penang and Johor remain the other key ports on the Malaysian side of the Straits. Penang's facilities have been expanded and the port has an important trade with Sumatra, together with an important export-processing zone. The development of a significant manufacturing base on the island as part of Malaysia's EOI strategy has meant a major boost to port activity. The port itself has some 13 general purpose, liquid and bulk berths, together with a container facility at North Butterworth. As with Port Klang, traditional exports of rubber and palm oil are combined with manufactured goods.

The growth of Johor Port, also known as Pasir Gudang, has been more specialised. The state as a whole has experienced an economic boom since the early 1980s, partly because of its proximity to Singapore and trading volumes, especially for container units, have increased rapidly over the last two decades. Six general berths and a range of facilities for dealing with hazardous chemicals and petroleum products have been developed. Port Dickson, trading around 6 million gwt per year is the fourth Malaysian port along the Straits. Originally founded to service the rubber plantation industry, it now functions as an important tanker terminal. Malacca today plays

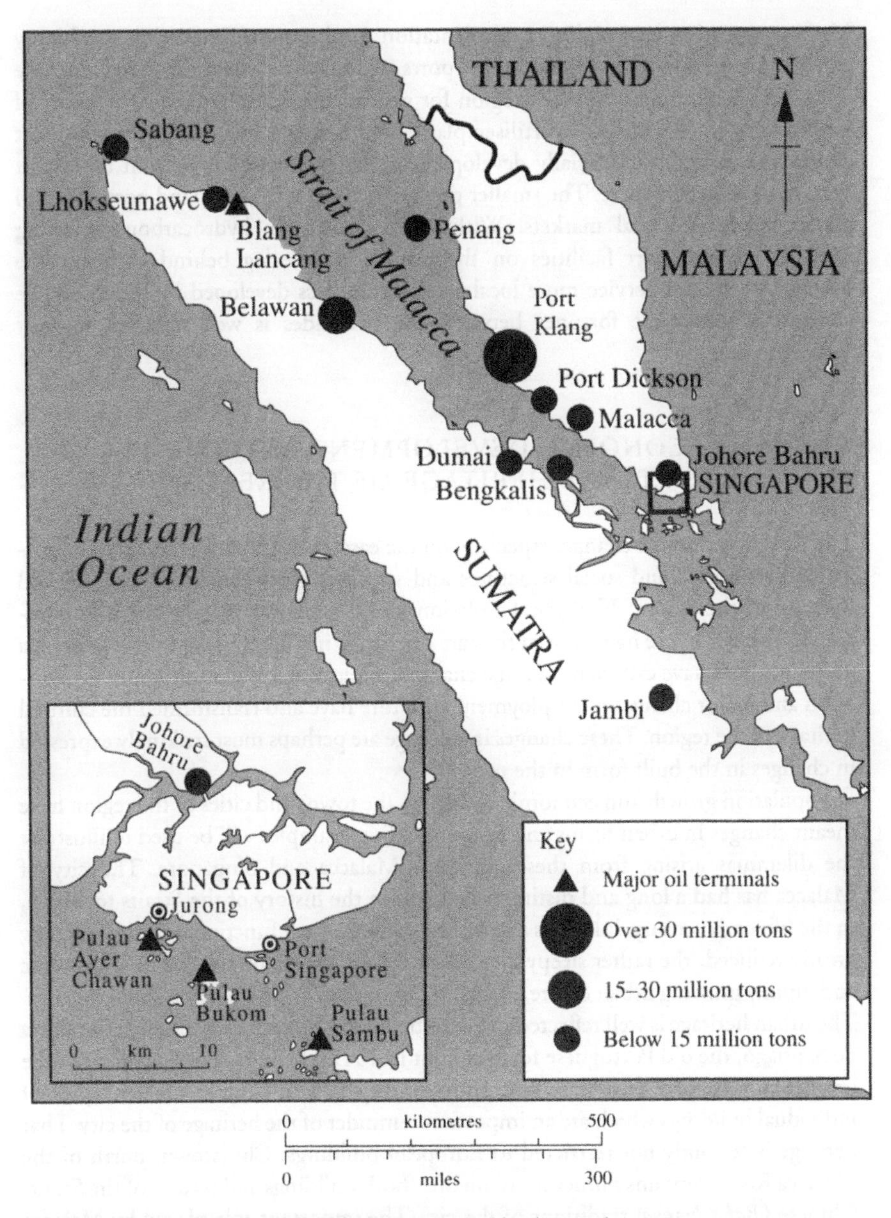

Figure 8.4 Major regional ports, mid-1990s (*source*: Lloyds, 1997).

only a very minor role as a port, a shadow of its former importance as a trading and entrepot centre.

Belawan is the most important general port on the Sumatran side, with trading volumes of around 19 million tons pa. It developed originally as the main port for

Medan, exporting large volumes of plantation products and importing machinery and consumables. Other important ports along the Sumatran coast include Lhokseumawe, serving the Aceh region for general cargo and also for the export of fertilisers from the ASEAN-Fertiliser plant in the region. The neighbouring port of Blang Lancang was specially developed for the export of LNG from the major Pertamina refineries there. The smaller ports of Dumai and the inland port of Jambi service essentially local markets. With exception of the hydrocarbon exporting facilities, general port facilities on the Sumatran side lag behind those of the Malaysian side and service more local markets and less developed hinterlands. The contrast in economic fortunes between the two sides is well reflected in port infrastructures.

ECONOMIC DEVELOPMENT AND THE CULTURAL HERITAGE OF THE REGION

The pace of economic change, especially on the eastern coast of the Straits, has transformed economic and social structures and wrought great changes in the cultural heritage of the region. If physical isolation and a generally inadequate communications infrastructure have slowed the rate of change in Sumatra, west coast Malaysia and Singapore have experienced huge changes. Education, migration, rapid urbanisation and major changes in employment structure have also transformed the cultural heritage of the region. These changes in heritage are perhaps most strikingly expressed in changes in the built form in the region.

Population growth and economic change in the towns and cities of the region have meant changes in urban forms and functions. Two examples will be used to illustrate the dilemmas arising from these changes—Malacca and Singapore. The city of Malacca has had a long and distinguished role in the history of the Straits to which, in the fifteenth century, it lent its name. Although the port function of the city is now greatly reduced, the rather sleepy city of the 1970s has been replaced by a dynamic and rapidly growing urban centre, as well as an important centre for regional tourism. The urban heritage is well reflected in a number of key buildings in the city. The Porta de Santiago, the old Portuguese fortress built in 1512, the Dutch town hall from the seventeenth century and the eighteenth century Christ Church are examples of individual buildings which are an important reminder of the heritage of the city. That heritage is certainly not restricted to European buildings. Chinatown, north of the Malacca River, contains numerous remnants (both buildings and trades) of the Straits Chinese (*Baba Nonya*) traditions of the city. The important role played by Malacca in the emergence of Malay nationalism is also an important part of the cultural heritage of the city.

The problems posed by the rapid physical transformation of the city have been well enumerated in Shaw and Jones (1997). Of prime concern is the fate of the old, Straits Chinese, core of the city. This part of Malacca, redolent with reminders of the past in its shophouses, residences and street plans, has been subject to major modification as the commodification of the *Baba Nonya* for tourist consumption proceeds apace. The waterfront area, traditionally a key part of the 'historic seascape', has been the subject of rapid change, with major land reclamation creating new sites for high-rise residential and commercial development. As Shaw and Jones (1997, 180) comment, 'as well as removing the visual element of the coastline together with the ambience created by the sea breezes, the smell of saltwater and sound of the sea, this reclamation drive is now impacting directly upon another aspect of Malacca's heritage, the historic Portuguese settlement'. Whether that historic fishing kampong can survive is likely to be an important test of the ability of the authorities to modernise the built form in a way which preserves key aspects of the cultural heritage of the city.

A second, much larger, example is Singapore itself. As was noted earlier in this chapter, the massive economic growth of the city, particularly after independence in 1965, was accompanied by the wholesale transformation of the built form. Facing huge problems of slum and squatter settlements, an inadequate infrastructure and a transport system largely neglected under colonial rule, the Urban Redevelopment Authority, a unitary authority established in 1974, spearheaded a massive campaign of housebuilding and slum clearance. In 1960, fewer than 10% of the population lived in public housing; by 1994, the proportion was 87%. Almost three-quarters of a million apartment buildings had been constructed in around 25 years (Powell, 1997, 86).

That such a huge redevelopment programme should impact on the historic built form of the city is hardly surprising. The housing programme meant the wholesale disappearance of kampong life: visiting the last village in Singapore, Kampong Wak Selat in 1993, was to witness the death of a uniquely Malay rural heritage. As Powell notes, the huge programme of urban renewal, undoubtedly an enormous achievement and, in many ways, a model of urban planning and management, inevitably led to the disappearance of familiar buildings, quarters and townscapes. From the mid-1980s, however, prompted partly by a fall in tourist visitors and revenues, something of a lobby for urban conservation began and the URA began to incorporate heritage arguments in its redevelopment plans.

It is possible to identify four major conservation areas in Singapore: the old core around the Singapore River, Chinatown, Kampong Glam and Little India. The redevelopment programmes for these regions incorporated the retention of many of the key architectural features of buildings and quarters whilst allowing for a range of functional changes to take place. The spectacle of the interiors of old shophouses being gutted and retaining only their ghostly facades was common from the early 1990s: for Powell (1997, 90), 'there was a tendency to over-conserve on the one-hand, "dressing-up" buildings in colours and details which are coarse and inaccurate, while on the other hand creating an overall blandness erasing the patina of age'. Other projects, most notably the redevelopment and expansion of the Raffles Hotel site, have sought to retain the form of past architecture whilst dramatically changing the function and context. As buildings disappear from the cityscape, a piece of the collective memory of the city and its people is lost.

Certainly the planning process is now much more sensitive to the cultural heritage of the city (Yeoh & Kong, 1995). The Concept Plan for Singapore, formulated in 1991, sees it as a Tropical City of Excellence, with a carefully planned and integrated set of 'heritage sites' to complement the 'green lungs' of the city elsewhere. It might be argued that much of the cultural conservation in the city represents pastiche rather than a properly contextualised past. A new 'Malay village' and the 'Tang Dynasty City' represent the invention of tradition, the stage at which the commodification of heritage for both internal and external consumption is complete (Shaw & Jones, 1997, 193).

Elsewhere in the Straits region similar pressures for change are evident. The form and functions of Georgetown, Penang, have been subject to dramatic changes as the nature of economic development has transformed economic and social structures in the city. Just as the colonial period has left its mark in urban fabric and form, so the impact of multinational capital and the requirements of the property market have been etched on the townscapes of the region. The differences between Malaysia and Sumatra, however, remain strong, with urban regeneration and redevelopment proceeding at a markedly slower pace despite high population growth. The much weaker development of the tourist economy in coastal Sumatra has undoubtedly contributed to the slower pace of redevelopment.

THE GROWTH OF TOURISM

The development of international tourism has had dramatic effects on the economies and built form of both Malaysia and Singapore since the late 1960s. When international arrivals in the region first started to pick up from the late 1960s, it was the appeal of low prices, excellent beaches and the climate that were important in attracting visitors. Penang (with its large resort area of Batu Ferringhi), Pangkor and Port Dickson have traditionally focused on both domestic and international tourism and have experienced a huge growth in the numbers of hotels and other infrastructure to support the industry. In addition to these traditional resorts, a number of important new destinations have emerged. Perhaps the most important of these are the island resorts.

The Langkawi group of islands represents one of the most important tourist developments along the Malaysian coast, with huge investments in international class hotels, the extension of the main airport to accommodate international arrivals and the development of new infrastructure to cater for tourists. Alongside this traditional area of expansion, the island resort has also tapped into the regional conference market: its annual Maritime and Aerospace Conference, for example, is the largest in the region. The island resort of Pangkor is a second example. Originally a small isolated fishing community, facilities have been dramatically upgraded over the last decade to accommodate the increased numbers of international tourists seeking high class accommodation in a tranquil environment. Next to the main island, the smaller island of Pangkor Laut has been developed for exclusive high class accommodation. Elsewhere on the Malaysian side of the Straits other, often tiny, islands have attracted the tourist dollar: Pulau Besar, for example, off Malacca, a very small, cramped island in terms of land for development, has already seen concerted efforts to develop an intensive tourist industry.

Many of these island resorts encapsulate both the positive and negative aspects of tourist expansion along the Straits. On the positive side, the revenues from tourism have made an important contribution to the economy, created employment and led to greater business opportunities for the population. But the economic benefits have to be set against the structure of tourism, particularly in the large international destinations. Control of the industry by large overseas companies can lead to the repatriation of profits and there is an argument that local benefits in terms of employment may be limited to the poorer paid positions in the industry. The impact of tourist development can also be significant in eroding or commodifying local culture.

It is perhaps in the environmental arena that problems may be most acute (Wong, 1996). The development of Langkawi, for example, has led to large-scale forest clearance for roads, infrastructure and hotels. Whilst much of the island is composed of granite, limestone hills dot parts of the island and these have, in places, been heavily over-quarried for construction materials in the building boom of the late 1980s. An important cement plant on the island has also drawn heavily on the limestone deposits. Given the range of pressures on the island, it is hardly surprising that the damage to the landscape has been so great. Increased erosion and sediment load in the coastal waters has also had a deleterious effect on the tourist industry and some of the off-shore corals, another important tourist attraction, have also experienced damage. Increased turbidity and waste discharge as a consequence of tourist development also threaten the shellfish industry in the off-shore waters.

The trend for increasing tourist development along the Malaysian coast, whilst slowing somewhat with the regional economic downturn in the mid-1990s, has impacted particularly on island resorts. Often very small islands, Pulau Besar off Malacca for example, are being developed as resort destinations for international tourists. Alongside these new island resorts, traditional resorts remain popular. Tanjung Bidara in Malacca, Port Dickson in Negri Sembilan and Banting in Selangor are good examples of beach resorts developed for a national rather than an international clientele. In some, much of the development has been unplanned and haphazard; elsewhere, state governments, acting through subsidiary companies, have sought to control the development process.

The contrast between tourist development on each side of the Straits is stark. If Singapore and the Malaysian coastal resorts represent mature, well developed and popular international tourist destinations, on the Sumatran side development has remained minimal. The reasons for this are not hard to elucidate. The physical environment is much more difficult from the development point of view. Extensive, mangrove-fringed coastal plains and mud flats create a very different environment, one which is both less attractive to international tourists and much more costly to develop. Investment in tourist infrastructure has been minimal. In part, this reflects the higher priority in government tourist budgets devoted to tourism elsewhere in Indonesia—the Mt Merapi district of central Java, for example, or the major tourist infrastructure on Bali. Political uncertainties in Aceh have also reduced potential tourist development there. It would seem likely that only small-scale eco-tourist development, catering for a specialised clientele, will emerge in the region.

Since the end of the Second World War, the Straits region has undergone probably the most intensive period of social and economic change in its history. Political independence, together with dramatic economic growth, has altered landscapes, lifestyles and the economic position of the region in the global economy. The fragility of that growth has perhaps been evident in recent years—with economic recession and boom and bust patterns of growth throughout the area—but the pace of overall change has been enormous. Post-war growth has perhaps accentuated regional differences between the western and eastern coasts of the Straits, with Malaysia and Singapore outstripping the pace of development on the Sumatran side. If economic diversity has characterised the last few decades in the region, the importance of political and economic cooperation between the regions of the Straits has undoubtedly grown. It is to the international dimension of the development of the Straits that the following chapter turns.

9

THE STRAITS: THE INTERNATIONAL DIMENSION

As one of the busiest stretches of water in the world, and with a location close to the heart of an important trading and industrial bloc, a range of international issues impinge on the use and management of the Straits. Both the coastal states involved and the many user states, most notably Japan, have huge strategic and economic interests in the Straits. Issues such as traffic management, policing and naval use have all been subject to problems over the definition in international law of the Straits. Such issues are considered in this chapter; international management of environmental issues is considered in Chapter 10.

THE QUESTION OF LEGAL STATUS

It was the Anglo-Dutch Treaty of 1824 which was to effectively partition the Straits between these two colonial powers. Prior to 1824, as Chapters 6 and 7 have shown, economic, political and social life criss-crossed the islands and inlets of the Straits and, at various times, powerful kingdoms existed which bridged the waters. The great kingdom of Srivijaya, as well as the sultanates of Malacca, Aceh and Johor, had all, at varous times, extended their political and economic control across both Sumatra and Malaya. But the treaty, which also gave the Dutch control of the Nauna, Anamba and Tambelan islands, clearly divided political control along the line of the Straits. As Leifer (1978, 13) comments, 'the colonial determination of 1824 was instrumental in ordaining post-colonial political boundaries and indeed, made possible the consolidation of the Netherlands East Indies'. Since 1824, it has only been during part of the Japanese Occupation period, from April 1942 to May 1943, that a single administration was responsible for large parts of both Sumatra and Malaya. When Indonesia declared its independence, Sumatra was an integral part of the national territory.

Three states currently retain important interests as they abut onto the Straits—Malaysia, Indonesia and Singapore. The first two are regarded as having territorial waters in the Straits themselves: they regard parts of the Straits as belonging to their national territories as well as having important economic and strategic interests. Singapore does not have any territorial claim to the waters of the Straits of Malacca, but has important commercial and strategic interests given that they control and lay claim to the Straits of Singapore into which the Straits of Malacca flow. All three states have a clear economic interest in keeping the Straits open.

There are a range of legal definitions relating to the oceans which have an important bearing on the status and management of the Straits by the three coastal states. Malaysia and Indonesia have clear, internationally recognised territorial claims to most, if not all, of the waters of the Straits. The extent of territorial claim has depended on international agreements relating to the extension of off-shore rights. Historically, three nautical miles off-shore was regarded as the territorial extension of coastal states; under that regime there would remain in the Straits a corridor of varying width classified as the 'high seas' where no interference with free passage could be countenanced. For both Malaysia and Indonesia, the existence of this narrow strip was a cause of concern, especially given the weight of traffic, often carrying hazardous goods, transiting the Straits. In 1957, Indonesia extended its territorial claim to 12 miles off-shore; Malaysia, in following suit, would have clearly compromised the existence of the 'high seas' corridor since, in places, the Straits would be entirely placed within one or other of the coastal states.

The third UNCLOS (United Nations Law of the Seas Conference) meeting during the 1970s sought, amongst a range of other issues, to clarify rights of passage and management in Straits that were narrower than 24 miles, twice the 12-mile territorial extension which was agreed on at UNCLOS 3. One of the solutions adopted was to seek to create a special category of rights of 'transit passage'. These would create rights akin to those applying on the 'high seas' for ships transiting through international straits whose waters, under the 12-mile territorial extension, might be regarded as 'sovereign' by the abutting coastal state.

The creation of this compromise position was complex and difficult. It allowed for the interests of both coastal and user states to be recognised. The interests of user states were both commercial and strategic. In the superpower environment of the 1970s, both the United States and the Soviet Union were especially anxious not to compromise free passage of their ships through the Straits and, together with a number of other users, especially Japan, free, untrammeled passage was seen as essential for economic reasons. The position in other straits such as the Straits of Gibraltar, Straits of Hormuz or the Straits of Messina posed similar problems of international maritime law. For the coastal states—Malaysia and Indonesia—a key consideration was their ability to manage the increasingly dense traffic through the Straits, to protect the Straits from pollution, and to manage their in-shore and off-shore fishing industries.

The UNCLOS Convention, which, after ratification, entered into force in November 1994, sought then to establish clear 'transit passage' rights to protect both maritime powers and coastal states. Transit passage, note Smith and Roach (1996, 289), 'is defined as the exercise of freedom of navigation and overflight solely for the purpose of continuous and expeditious transit in the normal modes of operation utilised by ships and aircraft'. The convention is also explicit in emphasising that transit passage cannot be suspended by coastal states for any reason.

The establishment of transit passage was crucial in the elaboration of the new convention. As Van Dyke (1997, 329) emphasises, 'the major maritime powers made it quite clear that they were willing to accept the enlargement of the 'territory' of the coastal states . . . if and only if freedom of movement for both commercial and military purposes was not jeopardised'. The replacement of 'innocent passage' by 'transit passage', combined with a territorial extension was the outcome of a long and often difficult set of negotiations. What, then, are the implications of the UNCLOS Convention for the Straits of Malacca?

A first point is that, any ships travelling through the Straits must abide by accepted international regulations; the basic standards laid down by the International Maritime Organisation (IMO) have to be obeyed. Thus, notes Gold (undated, 4), 'although the coastal states may legislate for passing vessels in safety and anti-pollution areas, they may only apply internationally agreed standards in its legislation'. Second, whilst there is nothing in the transit passage regime which makes environmental and navigational legislation impossible to apply, there is little doubt that it is not as easy to apply and develop these as it is for territorial waters. Thus, purely on pragmatic grounds, the regime for the coastal states in the Straits is not as favourable as it would have been had they been able to successfully press their claim for a clear 12-mile territorial limit.

Any legislation that the coastal states might wish to apply has thus to be developed either through international treaty or, more usually, through the IMO. This procedure is designed to prevent the application of a series of divergent unilateral sets of legislation being adopted by different states. The issue of whether, for example, Malaysia or Indonesia can require payment for, say, environmental damage is unclear under the UNCLOS Convention. Neither does the convention give coastal states an unequivocal right to know what materials are being transited through the Straits, even where they may be carrying potentially hazardous goods.

In both the Malacca and Singapore Straits much has been accomplished in the area of navigation and traffic management despite the legal difficulties outlined above. The development and adoption of a Traffic Separation Scheme (TSS) exemplifies cooperation between the coastal states. The extension of the TSS has been proposed to connect those operating in One Fathom Bank and the Singapore Straits scheme. This would provide a TSS in the Straits from 3″ 10′ N to its western entrance. Other examples of cooperation include attempts to suppress piracy, as noted in Chapter 8, and to develop anti-pollution measures in the region, measures described in Chapter 10. Together with other forms of traffic routes and the establishment of gazetted in-shore zones, they represent ways in which safety can be improved. Discussions over the development of pilotage schemes have concluded that, although they may increase safety, there would be major problems over the funding and jurisdiction of pilotage.

COOPERATION WITH USER STATES

Given the enormous strategic and commercial importance of the Straits, it is perhaps a little surprising that user state interest in the management of the waters has been

Table 9.1 Projects funded by the Malacca Straits Council 1968–1994

Project	*Period*	*Cost (million yen)*
Survey and chart work	1968–1982	1,783
Tidal studies	1976–1979	646
Removal of shipwrecks	1972–1978	1,435
Dredging work	1979	1,001
Navigational aids	1970–1994	4,026

Source: Adapted from Ono (1997, 246).

rather limited. In the debates over UNCLOS 3 and the formulation of the UNCLOS Convention, the interests of the major maritime and naval powers were largely restricted to ensuring that as few obstacles to free navigation as possible were enshrined in the convention. The net result—the regime of transit passage—might be regarded as the less desirable outcome from the Malaysian or Indonesian perspective.

Japan is the only user state to provide major technical assistance and funding towards the management of the Straits. That, of course, recognises the enormous interest Japan has in maintaining smooth passage for its hydrocarbon supplies through the Straits. As was noted in the preceding chapter, taking the alternative route through the Sunda Straits would add considerably to energy transport costs. Japan played an important part in establishing the Malacca Straits Council in 1968, a non-governmental consultative body, funded by Japan, which was designed to coordinate the efforts of the three coastal states to manage navigation and environmental issues in the Straits.

As Table 9.1 shows, Japan has also been able to provide technical and financial support for a number of navigational measures in the Straits, including hydrographic surveys, dredging assistance and help with navigational aids. In addition, Japan, under the auspices of the Malacca Straits Council, has established a revolving fund to help coastal states deal with the costs of clearing up oil pollution. Under the fund, set up in 1981 and managed for consecutive five-year periods, by Malaysia, Singapore and Indonesia, funds are made available rapidly to coastal states to combat oil spillages after incidents in the Straits. When insurance and salvage monies are paid out, the amounts advanced are returned to the fund for use at a later date.

ECONOMIC COOPERATION IN THE STRAITS

There has been a growing literature on the importance of economic complementarity and synergy amongst the countries of the Association of South East Asian Nations (ASEAN) which, together with the increasingly significant development of trade blocs (the ASEAN Free Trade Area, AFTA, was declared in 1992), provides an important context for economic growth in the region. With Singapore, the powerhouse economy in the region, at the southern end of the Straits, it is perhaps not surprising that it has been pivotal in looking for resource and product complementarity in its economic links with other parts of the region. Perhaps the best example of these international economic links has been in the development and extension of the growth triangle concept which has been viewed as a potentially valuable means for widening and deepening economic development. The first growth triangle, actively developed from the early 1990s, links Singapore, Johor and Riau province in Indonesia. That triangle has now been joined by a second, still at the formative stage, the so-called Northern Growth Triangle, linking northern Sumatra with Penang/Kedah/Perlis and southern Thailand.

The emergence of the growth triangle as an organising economic concept has been linked with the theme of much greater cooperation between the ASEAN members. ASEAN, founded in 1967, has been steadily developing its cooperative programmes and extending the already high degree of political cooperation into the economic sphere. Whilst it remains very much a pragmatic organisation, less structured than counterparts such as the EU, it is, nevertheless, increasingly providing an important means for driving trading and other forms of economic cooperation in the region. As a forum for resolving both political and economic issues, its role is likely to be enhanced in the future.

The emergence of the growth triangle owed much to the economic planning of Singapore in the late 1980s. As was noted in the preceding chapter, the city state had developed rapidly from the mid-1970s on the back of a strategy of multinational investment, attracted by low wage rates and attractive inward investment conditions. Overall growth rates, however, began to suffer in the mid-1980s for two reasons. First, wage rates in Singapore began inexorably to rise as the levels of education and expectations rose. Second, other states in the region—Malaysia and Indonesia, for example—began to secure more inward investment as part of their own development strategies (Rodan, 1987).

Faced with this problem, the Singapore government launched a twin policy of industrial change. First, there were major investments in high-tech, high value-added industry, a so-called Second Industrial Revolution focused on, for example, biotechnology, electronics and computer software. Second, there was a strategy of shifting low wage/low value-added production away from the state by encouraging such investment by Singapore companies into countries where the economic conditions were more suited to such a focus. As Perry (1998, 102) noted, 'those processes are said to be reflected in the *moving out* of low value assembly and manufacturing work and the *moving up* to a greater concentration of sophisticated production activities, design and development and other business service functions'.

From the mid-1980s, then, Singapore planners began to actively develop this strategy of industrial shift to take advantage of changes in the labour market and regional economy, and to look for complementarity and synergy between Singapore and other nearby production sites. The development of the growth triangle between Johor, Riau and Singapore itself was a key outcome. The development of greater synergies between these three nodes predated the formal establishment of a growth triangle by several years. Thus Johor had emerged as a key growth point in the emerging Malaysian economy and had, in any case, long supplied Singapore with both resources (notably water) and labour. A range of projects in communications (the second causeway for example, linking Singapore to Malaysia) and tourism further meshed the two nodes together.

The Riau islands, especially Batam Island, the closest Indonesian island to Singapore, had seen considerable investment by the Indonesian government from the late 1960s, focused primarily on developing its role in the oil bunkering and refining industry. The Batam Industrial Development Authority was given an enhanced role from the late 1970s and, from the mid-1980s, greater encouragement was given to attracting inward investment to Batam. The establishment of industrial estates coupled with real estate opportunities in executive housing and tourist accommodation were linked with rapidly growing investment by Singapore on the island. The 1989 Memorandum of Understanding between Singapore and Indonesia, 'unleashed a rapid expansion of Singapore's investment in Indonesia' (Perry, 1998, 93) and Batam Island became a prime destination for investment capital. The creation of the Batamindo Industrial Park presaged a major expansion of interest, with Singapore investors anxious to take advantage of the low wage rates, low social overheads and cheap land on Batam. Annual exports from Batam rose from US$44 million in 1988 to over US$3,000 million in 1996 (Perry, 1998, 94).

The place of Johor in this triangle of growth can perhaps be viewed as mid-way between the two. Unlike Batam where the chief attraction is cheap labour, wage rates and land costs are higher, though still well below those of Singapore. Inward investment has thus focused on mid-range products and in the leisure and retailing sector. But both Batam and Johor fulfil a role as a labour supply for Singapore. Singapore retains the key research and development functions and continues to be a major source of investment capital. Whilst there remains a degree of complementarity, both economic and political, between the three nodes, the growth triangle is likely to remain an effective part of economic strategy in the region. Critics have argued that links of dominance and dependency have been formalised as a result of the triangle and that the role of Singapore and the benefits to it have been too high. However, whilst the economic performance of the triangle remains positive, it is unlikely to be altered formally; only the changing nature of capital and labour in the region is likely to end its role.

The proposed growth triangle in the north of the Straits is much less advanced (Kumar & Siddique, 1994). Whilst flows of resources, capital and migrants have been especially important across the Straits in recent decades, only in the last few years has the idea of extending the growth triangle concept to this region been developed. The Northern Growth Triangle is designed to include Penang, together with the north Malaysian states of Kedah, Perlis and Pahang, the Indonesian provinces of Aceh and Sumatera Utara and the southern states of Thailand.

The triangle perhaps faces more difficulties in developing than its southern counterpart. On the positive side there are economic synergies; thus the presence of an important plantation sector may provide scope for cooperation in production and marketing across the states, whilst the triangle may aim to be self-sufficient in foodstuffs. The desire to decentralise from Java and the Bangkok district, respectively, might also encourage the Indonesian and Thai governments to develop the triangle. But the infrastructural obstacles are significant, particularly in Sumatra, and the triangle does not have the advantage of geographical proximity that its southern counterpart has. Political obstacles may also exist. Whilst Penang is an obvious hub, tensions between Penang and Kuala Lumpur may slow progress, whilst on the Sumatran side, political instability in Aceh is chronic and poses difficulties in developing cooperative arrangements across the Straits. The presence of large numbers of refugees from Aceh in Malaysia has periodically been a cause of political tension, particularly since the economic recession of the mid-1990s.

10

ENVIRONMENTAL ISSUES AND CHALLENGES

The Straits of Malacca are an important regional sink for a range of land-based organic and non-organic materials that find their way into the system via rivers, drains and canals. The extent and quality of these materials depends very much on the type of economic activities and land use in the littoral states. In addition, the level of control and management exercised by those littoral states in reducing the discharge of wastes and pollutants from those economic activities into the rivers and coasts is an important factor influencing the quality of the marine environment of the Straits. This chapter, focusing on environmental issues in the Straits, examines the nature of point and non-point sources of pollution, particularly those of anthropogenic origin. While land-based pollution is in general more serious, sea-based pollution due to infrequent but serious tanker collisions, frequent tanker bilge discharges and oil leakage from vessels is also important in affecting the ecology of the Straits. Efforts made by the littoral states individually, regionally and through international collaboration to manage the environment of the Straits will also be considered. In addition, the role of user states, international organisations and the international community before and after the UNCED summit in 1992, and the impact of Chapter 17 of Agenda 21 will be evaluated.

REGIONAL ENVIRONMENTAL ISSUES

It is misleading to consider environmental problems as resulting solely from the pressures of economic development in the Straits. Given the wet climate and high intensity rainfall, as well as the presence of well-weathered soils, natural erosion has been taking place over a long time on both sides of the Straits. Eroded sand, silt and mud have been flushed out into the sea episodically by high intensity storms and floods. In fact, as Chapter 2 suggested, surface erosion might well have been much greater when the Sunda land levels were higher. Over time, as sea levels rose and sediment transport to the coast continued, the broad plains in eastern Sumatra covering some 88,000 sq km or 18% of the island's total area (Burbridge, 1988), and the coastal flatlands of western Peninsular Malaysia began to take their present form. The deposition of colluvial-alluvial and finer materials into the coastal waters of the

Straits of Malacca has thus produced large tracts of mangroves and peat swampland, leading to a progressive narrowing of the Straits, constriction of the deeper channels and an overall reduction in depth of certain parts of the Straits (Keller & Richards, 1967).

The natural processes of erosion, transport and deposition are particularly effective in the climatic environment of the Straits region, even when total forest cover is present. Fournier (1960) pointed to the high rates of erosion under natural conditions in Malaya, rates of between 2,000–3,000 metric tonnes km^2 yr^{-1}; Douglas (1967) argued that the total loads carried by streams under natural, undisturbed conditions may be fairly high, up to 100 cu m/sq km/yr. High rates of soil detachment by rain drops of high intensities (>100 mm hr^{-1} is not uncommon) and surface wash and gullying have also been noted on a range of slopes. Several studies of lowland forests, carried out either on the basis of catchments or erosion plots, showed significant sediment removal (Leigh, 1978; Peh, 1978). The extent to which variables such as slope value, topography and type of forest cover influence erosion rates has not yet been fully studied or explained. However, information from current research points to consistently high sediment removal, even from forested areas.

Human impacts through economic development in Peninsular Malaysia, Singapore and Sumatra have been evident over a long time period; what has perhaps been most striking in environmental terms has been the greatly accelerated impacts in recent decades. Economic development in all the three territories has historical roots, in particular during the colonial era. The major towns, ports, road and rail networks that accompanied mineral resource exploitation and the expansion of plantation agriculture on both sides of the Straits largely emerged in the colonial period. As Chapters 7 and 8 showed, by the early twentieth century much of Peninsular Malaysia, especially its western belt, Singapore, and to a lesser extent, the provinces of Aceh, North Sumatra and Riau, had been heavily transformed by the development process.

In Peninsular Malaysia, particularly in the states of Perak and Selangor, mining of tin, which started in the late eighteenth century, had caused large areas of the once forested areas to be cleared and excavated, leaving a severely scarred landscape (see Chapter 4). Water from rivers was drawn to feed hydraulic pumps in order to break walls of weathered materials down, or used to wash the materials in troughs (*palongs*) so as to allow the heavier tin ores to be separated from sand and silt. Whilst the sand and silt portions were usually collected as tailings, much of the slurry found its way into the natural watercourses, thus polluting streams and rivers. The introduction and expansion of cash crop agriculture such as rubber, oil palm and cocoa has resulted in large areas of forest in Peninsular Malaysia and Sumatra being converted to plantations. The method of operation was poorly regulated in the early years—large areas of forests on undulating and even hilly lands were cleared and dry vegetation burned, and the erosion of surface materials continued unchecked.

Given the high intensity storms that characterise the rainfall regime, large-scale land development and a lack of success in implementing a plethora of environmental by-laws, it is not surprising that outputs of eroded sediments have been extremely high, especially in areas of high topography. Quantitative measurements made by

Shallow (1956) of two catchments with different proportions of natural vegetation in the Cameron Highlands, showed that a catchment that has 36% of its forest cover disturbed produced five times more sediment (103 m^3 km^3 y^{-1}) than a similar sized catchment that was almost totally forested (21.1 m^3 km^2 y^{-1}). Daniel and Kulasingam's study (1974) from the same area examined three catchments with differing land uses, namely forest, tea and vegetables. Sediment yields of 25 m^3 km^2 y^{-1}, 488 m^3 km^2 y^{-1} and 732 m^3 km^2 y^{-1} were found to be produced by the three catchments, respectively. These high sediment yields as a result of agricultural activities on steep slopes have resulted in the rapid deposition of materials into the Ringlet Hydroelectric Dam system, threatening its life span and economic viability. In basins where river flow is unimpeded, these large sediment quantities are discharged direct to the coast.

Amongst the major river systems flowing into the Straits it has been the Klang River, and its tributary the Gombak, that has been the subject of most research. It is within the larger Klang valley that some of Malaysia's intensive urban development in the so-called Kuala Lumpur–Port Klang developed corridor is located. While the river has always contained high suspended load due to tin mining, recent urban and industrial development has made the river one of the most polluted in the country. Like the Klang, other rivers in Selangor are similarly affected. In fact, samples of suspended sediments in the rivers in Selangor, in particular, at the mouth of Selangor River and at Sungai Buloh in 1992, showed high values of 2,296 ppm and 2,008 ppm, respectively (Dow, 1997). Another major river in west Peninsular Malaysia that brings high sediment concentrations to the coastal waters is the Perak River, where extensive tin mining activities are largely responsible for the discharges.

Subsequent river discharge into the Straits has had a direct influence on their marine ecology through swamp expansion, sea grass growth and coral life. Coral reefs in the west coast of the peninsula are severely affected by sediments flowing off the land as a result of land clearing and industrialisation (Chua *et al*, 1997) due to an increase in turbidity and a decrease in water transparency, limiting the penetration of sunlight. This is especially the case during peak flood flows when large sediment plumes visibly spread out into the Straits from estuaries, often vividly captured in satellite pictures. Transparency values of less than 10 m are common in estuaries along the coasts of Malaysia and Sumatra, while in the open sea they may vary from 10 m to 30 m. While reef development in some 241 islands off the coastal zone of west Malaysia may not be as varied or extensive as on the east coast, it is nevertheless important for marine life. In Singapore, some reefs have been affected by turbidity associated with extensive land reclamation projects (Chou *et al*, 1994).

The construction of impoundments and dams on both sides of the Straits has reduced river discharge and, in theory, the sediment output to the coast. In Peninsular Malaysia major drainage control programmes (see Figure 3.1) have resulted in some 12% of the natural runoff being intercepted. This value will rise to 51% when all the planned and proposed dam projects are completed (Khairulmaini, 1994); most of these are found in the western half of the peninsula with river systems draining into the Straits. Despite this potential reduction in discharge, sediment output continues to be high. On the Sumatran side, the river runoff is less regulated. However, there are large dams that have been built, notably the Asahan Dam on the Asahan River for hydro-electricity generation purposes. Nevertheless, since the proportion of river discharge that has been regulated is small relative to the total annual discharge, the influence of rivers in transporting sediments to the coasts remains significant.

The river systems draining the slopes of the Barisan Range carry substantial quantities of sediment which are deposited within the channel, on the coast and in the Straits. In Sumatra, pollen analysis has shown that humans have been clearing the forest in the upland areas for some 7,000 years or more. Shifting cultivation did allow forest regeneration but from about 2,000 BP permanent clearings began to be made (Flenley, 1988). In fact, the wide coastal plains of eastern Sumatra suggest that this coastal accretion is sustained by continued sedimentation (Verstappen, 1973) by major rivers like the Rokan, Siak, Kampar and Indragiri. The east coast of Sumatra has been built up into the Straits largely through deposition of the considerable sediment load of these streams (van Bemmelen, 1949). In fact, Sumatra has the greatest influence on the sediments of the Straits of Malacca, although Malaysian rivers contribute large quantities of material (Keller & Richards, 1967).

It is difficult to quantify the amount of sediments brought down by rivers in Sumatra as sediment studies are few. Instantaneous sediment measurements in southern Sumatran rivers in recent years indicate values of 40 ppm to 500 ppm. In the downstream reaches of the river silt particles are smaller than 0.05 mm. Some results suggest that annually some 90 m of coastal land is added in eastern Sumatra. This has caused some former off-shore islands to be absorbed, and existing swampy islands such as Rupat, Bengkalis, Padang, and Rantau, which are still separated from Sumatra by narrow sounds, will be connected to it in the near future (Donner, 1987). Although new coastal land is added due to accretion, sedimentation has also threatened coastal ports (Pelzer, 1968; Verstappen, 1964). A good example is Bagan Siapi-api, one of Indonesia's largest fishing ports. It has been pushed some distance inland and access through its rapidly silting navigational channels is becoming increasingly difficult (Chua *et al*, 1997). In the west coast of Peninsular Malaysia, the annual dredging of sediments at Port Klang alone amounts to some 226,000 cu m of soil and maintenance dredging of fishing ports of Kuala Perlis and Kuala Kedah annually remove some 50,000 cu m (Cho, 1995). Dredging activities have had to be carried out in Penang Port in recent years.

COASTAL EROSION AND LAND RECLAMATION

Coastal erosion and land reclamation, by virtue of their taking place at the interface between land and sea, contribute sediments directly into the sea and affect the immediate marine environment. The problem of coastal erosion is especially serious along certain stretches of the Malaysian coast; the problem is much less serious on the Sumatran side given the extensive mangroves and the low energy environment, except in the northern parts of the province of Aceh.

In Malaysia, the seriousness of coastal erosion led to the commissioning of a major enquiry by the Malaysian government in 1992. The study found that out of the 1,110 km length of the coastline in Peninsular Malaysia bordering the straits, 145 km showed serious erosion that threatened, within the short term, shore-based facilities; 246 km showed significant erosion where the facilities were expected to be in danger within five to ten years if remedial action was not taken; and 975 km showed acceptable erosion rates. The average rate of shoreline retreat ranges from less than 1 m to more than 10 m per year. Of the 145 km in the first category of serious erosion problem, about 131 km occur at 43 locations along the coastline of Peninsular Malaysia and 36 of these face the Straits of Malacca (Abdullah, 1992).

Coastal erosion is a complex process but several factors can contribute to it, such as exposure to storms, strong waves and currents, or the loss of mangroves which hitherto acted as a stabilising influence by absorbing the impact of currents and waves and by acting as an effective sediment trap. Other causes include the curtailment of sediment supply as a result of river regulation, reductions in fresh water output to mangroves so essential to the wellbeing of the mangrove ecosystem, and mining of sand along the coast. These are relevant in the consideration of coastal erosion on both sides of the straits.

Sediment output from coastal erosion exacerbates the problem of turbidity and siltation due to river discharge. The problem can also be further compounded by land reclamation activities which are not carefully controlled. In west Malaysia, as available land for development is becoming scarce and expensive, coastal areas lend themselves to government-sponsored reclamation. Many places along the west coast have been reclaimed for the construction of new townships or the extension of existing towns. Malacca town, for example, has expanded through the construction of shopping complexes within reclaimed coastal land in Kota Melaka.

Singapore has over the past three decades added some 10%, or approximately 83 km sq of land, through land reclamation, such that there are now very few stretches of coastline on the island that retain their natural indented nature. Current land reclamation projects include amalgamating four separate southern islands southwest of Singapore into one single large piece of land for a petrochemical and chemical industrial complex. While the reclamation is necessary to create space for housing, port and airport facilities, industrial sites and land banks, the siltation process has inevitably resulted in some harm to coral reef growth as well as to the coastal habitat of the smaller islands and increased sedimentation rates in Singapore waters (Low & Chou, 1994).

Apart from inorganic soil materials transported into the Straits of Malacca as a result of soil erosion and sediment transport, and from coastal erosion and reclamation, the Straits also receive inputs of natural discharge of organic seepage from peat swamp forests and solutes from rock weathering. Inputs of tannin and other acidic discharges from the decomposition of organic matter in swamp forest find their way as dark coloured water flowing into the sea, easily detected on satellite pictures. Eastern Sumatra, with a natural peat area of 7.3 to 9.7 million ha (Andriesse, 1974) or about 25% of all tropical peat lands (Driessen, 1978), is naturally the main contributor of organic input into the western waters of the Straits. Studies using remote sensing techniques on organic inputs into the coastal waters of Johore show clearly the input of this kind of discharge in contrast to discharges from naturally weathered rocks (Nichol, 1993). What is interesting is that organic water continues to leak into the sea long after the peat areas have been reclaimed or even developed.

Nutrients also come from other sources, such as the weathering of rocks, precipitation and smoke. Chemical weathering under natural conditions releases several important elements such as calcium, magnesium, potassium, sodium, silicon, iron, aluminium and phosphorus. The types and quantities of these elements produced depend very much on the type of parent materials undergoing weathering. However, given a forested environment and relatively thick layer of saprolite, closed or nearly closed nutrient cycles may be more widespread as confirmed by several studies in Malaysia (Burnham, 1989; Hamdan & Burnham, 1996). But in exposed rocks nutrients may find their way directly into streams.

WASTE DISCHARGE

In the regions abutting the Straits, a range of settlements, agricultural projects and factories of all sizes have contributed a number of waste discharges including garbage, waste water and sewage from households; organic wastes, fertilizers, herbicides and insecticides from agricultural activities; chemicals, toxic (heavy metals) and non-toxic wastes from industries; and storm-water runoff from urban areas. Until recently all these discharges found their way into river systems and, ultimately, into the Straits.

Direct untreated waste from human settlements has been discharged into the rivers of the region for many centuries and, to some extent, this continues through to the present in many of the less developed cities and rural settlements on both sides of the Straits. Domestic sewage comprising excreta and wastewater from household facilities, collectively known as sullage, has been increasing rapidly with population growth and expansion of settlements. The effects of such waste discharge into rivers and the coastal waters are the high BOD (Biological Oxygen Demand) loadings. It was estimated in 1989 that the BOD loading from domestic sewage discharge in the coastal areas of Indonesia, Malaysia and Singapore was 5,014 tons per day (see Table 10.1).

Much of the sullage comes from the water that is supplied to homes and, in the design of treatment wastes, a value of 90% of water consumption or 210 l/cap/day has been adopted for the Klang Master Plan (an overall management strategy for the Klang River) to account for this. A similar value of 225 l/cap/day (SIRIM, 1989) of waste water has been proposed for Malaysia as a whole. However, the waste water is expected to increase over time as the consumption of water increases. In 1995, the waste water generated was expected to increase to 250 l/cap/day and for 2000 in the Federal Territory, sewage flows are projected to be 180 l/cap/day for low cost housing areas and 256 l/cap/day for a mixed medium-high cost housing area (Ariyathavaratnam, 1989). Waste water generated from commercial areas can be

Table 10.1 Estimated BOD loading from sewage discharges in coastal areas bordering the Straits

Parameter	*Year*	*Indonesia*	*Malaysia*	*Singapore*
National population (million	1989	184.6	20.14	2.7
Total BOD load (ton)		9,230	870	324
Coastal population (million)		110.76	12.8	2.7
% of national population		65	60	100
Daily BOD load in coastal area generated (ton)		5,538	565	324
Primary treatment (%)		60	70	10
Secondary treatment (%)		0	5	90
Daily BOD removal (ton)		997	144	272
Residual daily BOD disposal (ton)		4,541	421	52
National population (million)	2000	222	20.9	2.9
Total BOD load (ton)		11,100	1,045	348
Coastal population (million)		133.2	13.58	2.9
% national population		60	65	100
Daily BOD load in coastal area generated (ton)		6,600	679	348
Daily BOD removal (ton)		1,199	173	313
Residual daily BOD disposal (ton)		5,461	506	35

Source: Koe and Aziz (1995).

much higher and an estimate of 460 l/cap/day was used in Seberang Perai (Majlis Perbandaran Seberang Perai (Seberang Perai City Council), 1981).

About 80% of organic pollution in Malaysia in the late 1980s was due to domestic sewage (Department of Environment, 1989). Sewage pollution has persisted because of the lack of financial support to provide a more comprehensive sewerage system. Being a highly capital intensive investment, commitments to sewerage projects have been extremely slow. Only recent awareness of its effects on the tourist industry and water resources seem to have created an interest in taking sewage management seriously (Suki, 1993).

A range of sewage treatment plants are in use in Malaysia and these include oxidation ponds, aerated ponds, oxidation ditches and activated processes, imhoff tanks and septic tanks, rotating biological filters, trickling biological filters and other fixed media systems. Problems of maintenance have resulted in lower efficiency in BOD removal, especially in the case of oxidation ponds, aerated ponds and oxidation ditches. For oxidation ponds, a BOD removal efficiency of 58.3% was noted, whilst for rotating biological contractors it was 6.7% and the SS removal efficiency was 51.6% (Akira, 1990). The poor performance of sewage treatment plants can result in the release of *Salmonella* into the environment. Yaziz (1981) indicated that the removal of *Salmonella* in sewage using primary settling ranged from 0–80% depending on the maintenance of the unit. The removal of *F. coliforms* was better: 81% for sedimentation tanks, 52% for the primary ponds, 92% for the secondary ponds and 99% for the whole pond system (Nik Fuad, 1988). There is evidence that the BOD of raw sewage entering a pond system has declined with time (Suki *et al*,

1987). The reasons are uncertain, but it could be due to a general increase in water consumption or infiltration (Suki, 1993).

A major Malaysian environmental quality report in 1988 (Department of Environment, 1989) clearly showed that sewage is an important source of organic pollutants and nitrogen compounds entering the rivers and, ultimately, the Straits. Studies in the Klang valley (Government of Malaysia & Asia Development Bank, 1987) indicate that the major polluter of the Klang River is domestic sewage. The effect of pollution by sewage was detected at the estuary and coastal waters off Port Klang. High *F. coliform* counts of 3 × 10 MPN/100 ml were detected in the river mouth, and at 10 km away counts of 30–50 MPN/100 ml were detected (Law, 1980; 1984). These counts were high considering the inactivation effect of high salinity waters on the *F. coliforms.* The effect of sewage pollution in the coastal waters off Port Dickson has also been investigated (Law & Azhar, 1985). The results of this study showed a high count at up to 5–6 km from the shore. Out of 17 sampling locations, only a few could be considered safe for recreational purposes. Measurements in 1995 of *E. coli* showed readings exceeding the Proposed Interim Standard of 100 MPN/100 ml in the coastal waters of Melaka and Selangor between March and May (Department of Environment, 1996), indicating that treatment of wastes had not improved.

In Singapore much of the sewage and sullage water generated by human activities is treated. Water consumption by the population, industry and commerce, shipping and government and statutory boards has increased over the years and in 1997 the total was 441 million cu m. In 1990, the total volume of waste water that was treated was 320 million, and this is expected to almost double by 2025 (Tan, 1992). The east coast of Sumatra had a population of about 11 million in 1994 and assuming that 60% of the sewage generated reaches streams and rivers that flow into the Straits of Malacca, total BOD loadings, COD, TN and TP are estimated at 167,292 tons, 381,425 tons, 74,278 tons and 7,361 tons per year, respectively (Chua *et al*, 1997).

Pollutants discharged from the major rivers flowing into the Straits from Peninsular Malaysia are varied. The Perak and Klang rivers have high values for most of the pollutants in terms of TSS, BOD, COD, lead and copper by reason of high concentrations of industry, intense development within the catchment and high population densities.

Solid wastes

Garbage production has been increasing over the years in Malaysia, Indonesia and Singapore, especially with large increases in population and a steady rise in the standard of living which results in corresponding increases in consumption. This is especially true of the larger cities and towns where the local and municipal governments have yet to grapple with the problem of waste management and disposal. In the past, waste collection was carried out by local authorities which was unsatisfactory in terms of investment and collection efficiency. Privatisation of solid waste management services has been introduced in many municipal council regions in

Malaysia with mixed results. It appears that privatised rubbish collection services in 17 municipalities showed that collection efficiency improved in 11 of the 17 (Lee, 1997).

Landfill is the most common method of disposal. Though there is a Recommended Code of Practice for Landfill Development and Management, very few operating landfills meet the engineering standards for sanitary landfilling that should include the installation of proper liners and leachate collection systems to prevent contamination of surface and ground water (Chua *et al*, 1997; Norhayati *et al*, 1994). In most cases open dumping of waste is practised. In Kuala Lumpur, landfill is the only method used for refuse disposal but sites are limited, while in Johor Bahru, perhaps because of convenience, garbage has been found dumped in swampy areas at the estuary of the Tebrau River (Lokman & Othman, 1991). The situation in cities like Aceh and Medan in eastern Sumatra is no better. In Singapore, 73% of the total refuse is disposed of in three incinerators at Ulu Pandan, Tuas and Senoko, while the rest is disposed of at the Lorong Halus dumping ground. A leachate treatment plant was constructed at Tampines in 1994, and a reclamation project of the foreshore area off Pulau Semakau as an off-shore landfill site started in 1995.

Agricultural wastes

Water pollution due to agricultural wastes has been a significant factor in the region, especially with the expansion of a range of agro-industrial processes linked to plantation production in both Malaysia and Sumatra (Jaafar & Harun, 1979). Agricultural wastes that are deposited in the Straits have come from two main agricultural activities—palm oil and rubber-processing factories. Along the west coast of Peninsular Malaysia there are 81 palm oil mills and 171 rubber-processing factories, and the total agro-based industries including tapioca processing and pineapple canning factories numbered some 375 in 1988, employing many thousands of people (Chua *et al*, 1997).

In Malaysia, with the increase in palm oil acreages in recent decades, the production of palm oil had increased fivefold from 92,000 tons in 1960, to 431,000 tons in 1970. By 1989, total production had gone up to 6.05 million tons, accounting for some 60% of total world production. This rapid increase in production necessitated the establishment of palm oil mills to process the raw materials. From just 10 mills in 1960 the number increased to over 250 in the early 1990s of which 85% were in the west coast states of the peninsula abutting the Straits. Untreated palm oil effluent has a high BOD and, on average, for every ton of fresh fruit bunch (FFB) of oil palm seeds processed, an effluent volume of 3.07 cu m per ton will be generated with a BOD of 31.3. The extremely pollutive nature of palm oil effluent is reflected in the fact that processing 60 tons of fresh palm oil fruits in one hour produces a daily BOD load equivalent of 900,000 persons. While there are other effects of oil palm waste discharge on the environment, none is more significant than its impact on the organic load in rivers and, subsequently, the Straits themselves.

In recent decades, then, palm oil wastes have become a significant environmental issue in the region. The significance of palm oil wastes is evident in their contribution to the total BOD load in Malaysian rivers. According to Ho (1987), oil palm mills contributed almost two-thirds of the BOD load generated in 1981, domestic sewage some one-quarter and rubber processing and other manufacturing industries most of the remainder. In 1975, the quantity of effluent produced by this sector alone amounted to over 3 million metric tons. In terms of its oxygen-depleting capacity determined by the BOD, oil palm waste has 100 times stronger effects than domestic sewage. Thus the total BOD load discharged into water courses was estimated at approximately 460,000 pounds per day or 210 metric tons per day. In terms of population equivalent, the BOD generated by the industry in 1980 was equivalent to a human population of approximately 10 million. By 1989 the production of effluent had increased to 15.2 million tons, which in terms of BOD demand was equivalent to a population of 22.3 million people, which is more than the total population of Malaysia (Ma *et al*, 1993).

The need for fresh water in the production process has also made palm oil waste of particular concern because most mills are located close to rivers for the cheap supply of water as well as for the convenient discharge of untreated wastes. Being highly organic in nature, the decomposition of palm oil wastes consumes large amounts of dissolved oxygen in the water. From a dissolved oxygen value of 8 ppm in unpolluted river water, oil palm wastes can result in the depletion of DO to a level of below 2 ppm. At this anaerobic level the river is no longer able to support fish and other aquatic life. The complete depletion of DO in rivers was not uncommon in the 1970s and 1980s, causing fish to die en masse. In addition, there was the release of noxious gases, particularly hydrogen sulphide, causing rivers to be malodorous and rendered useless for their principal users.

Legislative measures were introduced in the 1970s to try to curb pollution of rivers by oil palm waste discharge. A four-year programme to progressively reduce pollution from the oil palm industry was introduced. The promulgation of the Environmental Quality Act of 1974 provided for a comprehensive regulatory approach to the control of all kinds of industrial sources of pollution, as well as standards for the regulation of oil palm, rubber and other industrial effluents. Pragmatic regulatory approaches were incorporated in the Environmental Quality (Prescribed Premises) (Crude Palm Oil) Regulations, which came into force in November 1977. Under this regulation, the progressive reduction of pollution in the palm oil industry was envisaged over a four-year period: a 75% reduction in the pollution load by 1 July 1978, 90% by 1 July 1979, 95% by 1 July 1980 and 97.5% by 1 July 1981. At the same time, a four-generation set of effluent standards was defined by the Department of the Environment which was spelt out to companies seeking licences to establish and operate new palm oil mills. This standard was to be applied throughout Malaysia.

Beyond this regulatory approach, the siting of oil palm mills has now been controlled at the planning stage. Conditions must be met by companies before approval is given to them to set up factories. Further conditions attached to new licences include requirements for operators to set pollution control measures, and ensure the installation of pollution control and monitoring devices. A fees structure in respect of licences based on the levels of organic waste discharges from oil palm mills was established to encourage in-house installation of pollution reduction measures with the ultimate objective of reducing waste discharge into the environment. These rate charges distinguish between the ultimate modes of disposal of the treated or untreated wastes: ultimate disposal being permissible either into the watercourse or on land. Land disposal of the wastes takes the form of application of wastes as a substitute or supplement for fertiliser use. This is generally preferable to direct watercourse discharge, despite the higher BOD levels that may be tolerated. However, the corresponding permissible levels for watercourse disposal need to be significantly more stringent.

It is now in the interest of mills to tackle the problem of effluent discharge from their establishments. The fee structure has made it worthwhile for mills to institute some pollution reduction measures. An average-sized mill (20–20 metric tons capacity) which discharges effluent having a BOD concentration of 5,000 ppm will be required to pay an effluent-related licence fee of approximately RM4,500 for the first year. The above average-sized mill, if discharging raw effluent (i.e. effluent without any form of treatment) will be required to pay up to approximately RM140,000 for the first year, irrespective of the ultimate mode of disposal. Where effluent is neither discharged into a watercourse nor onto land, thus eliminating any environmental pollution problems, the fee payable is the minimum amount of RM150 pa.

Since the enactment of the Environmental Quality Regulations in 1978, several treatment systems have been developed by the palm oil industry in Malaysia including anaerobic, aerobic and facultative processes. More than 85% of the palm oil mills in Malaysia have now adopted the ponding system applying the anaerobic method for the treatment of the waste (Ma & Ong, 1987). In the close tank digester, biogas is produced which is harnessed for heat and electricity generation. In all these methods, the treated waste in the form of solids is used as fertilisers. Many researchers have shown the high plant nutrients found in raw and treated waste (Lim, 1987; Lim *et al*, 1984; Mohd Tayeb *et al*, 1987; Tam *et al*, 1982). It has been estimated that 15 million tons of waste would produce some M$87 million of fertiliser. Land application of the effluent is allowed if the effluent is less than 5000 mg/l. This 'recycling' of waste as nutrients for plants not only saves on costs of fertilisers, but gives benefits such as increased crop yields and improvements in soil properties (Mohd Tayeb *et al*, 1987). In the east coastal lowlands of Sumatra, North Sumatra and Riau Province are the main palm oil growing areas. The control of oil palm waste is not as well established and regulated as in Malaysia. There is little information about the extent of pollution reduction measures that have been instituted by mills in Sumatra, neither is there information on legal controls and implementation of environmental laws by the government.

Wastes from the rubber-processing industry are a second, significant, contributor to organic pollution in the Straits region. For one ton of SMR (Standard Malaysian Rubber) block produced, there is 20.5 cu m effluent containing 21.3 kg of BOD (Department of Environment, Malaysia, 1989). It was estimated in the late 1980s that about 100 million litres of effluent containing about 200 tons of BOD was being

Table 10.2 Chemical characteristics of effluent from two main types of rubber processing factory

Parameter	*Latex concentrate*	*SMR*
pH	3.7	5.7
Total solids	7,576	1,915
Suspended solids	182	237
BOD	3,192	1,747
COD	6,201	2,740
Total nitrogen	616	147
Ammoniacal nitrogen	401	66
Sulphate	1,610	–

Source: Zaid (1993, 139).

discharged daily from this industry. The present status of effluent discharge from rubber is not confined to raw rubber processing, but also comes from a variety of factories manufacturing rubber products. The concentration of such plants abutting the Straits has led to major pollution flows into the Straits waters.

Two types of natural rubber-processing plants can be distinguished—latex concentrate and the SMR. In 1986, there were about 38 latex concentrate and 143 SMR factories actively operating in Peninsular Malaysia. With an increase in the latex price, the number of latex factories increased to 84 in 1988 and the SMR factories decreased to 124 in 1990.

The two effluents produced in both types of factories have common characteristics as strong acids are used in the coagulation process. Thus, apart from high concentration of acids, the effluents also contain substantial amounts of organics and nitrogen. The effluent from latex concentrate factories contains high levels of sulphate which originate from sulphuric acid used in the coagulation of skim latex (Table 10.2). This high sulphate content in the effluent can cause problems in the biological anaerobic treatment system as high levels of hydrogen sulphide will be liberated to the environment and cause malodour problems.

Rubber processing factories are chiefly located in states abutting the Straits, especially in the Perak, Linggi and Muar river systems. Like palm oil effluents, compliance to four generation regulatory standards was made compulsory in the 1980s. The Rubber Research Institute of Malaysia, in keeping with the DOE's objectives, has its own housekeeping standards with regard to rubber processing factories; it endeavours to:

- Minimise the use of water during processing.
- Provide adequate systems of drains for removing spillage and effluent.
- Provide an efficient rubber trap with a minimum retention time of 12 hours for the recovery of un-coagulated latex and for mixing of the effluent.
- Provide deammoniation towers in the case of latex concentrate factories in order to reduce the level of free ammonia in the skim latex and also in the effluent.

- Minimise the use of sulphuric acid in skim latex coagulation for latex concentrate factories.

Several effluent treatment systems have been recommended by the Rubber Research Institute of Malaysia for factories which include the anaerobic/facultative ponding system, aerobic system, land application and enclosed anaerobic digestion.

In the rubber-related factories, the process of making rubber products involves the use of chemicals, water and latex compounds, which in turn produce industrial effluents that need to be treated before discharging into the environment. The effluents contain fairly high levels of solids, COD, BOD and nitrate, the last due to the use of calcium nitrate in glove and swimming cap manufacturing. Heavy metals like zinc and other metals like iron, magnesium and calcium are also found originating from the chemicals used. Treatment of these effluents varies from activated sludge processes to ponding and sand filtration or a combination of these. While 95% of all latex processing plants have suitable treatment systems for their effluents, the manufacturing sector still lags behind in this respect.

Other agricultural pollutants

Despite continued development in the industrial sector, agriculture still plays a significant part in the economy and livelihood of the peoples of the region. With large acreages of agricultural crops—whether for export or as domestic food crops—has come the widespread use of pesticides and fungicides. It was estimated that in 1980 pesticides sales in Malaysia alone amounted to R160 million, and this increased to R237 million in 1984 and to nearly R300 million in 1990.

With over one-third of a million hectares of agricultural land along the states bordering the Straits of Malacca, pesticides and fertilisers are used extensively. Three crops, however—rubber, oil palm and rice—account for 90% of the annual pesticide use. Seventy-five per cent of the pesticides are used for rubber and oil palm plantations. Insecticides are more commonly used on vegetable, rice and tobacco crops. Among the chemicals used in agriculture, herbicides account for nearly 80% of the total pesticide market in Malaysia, insecticides 15%, while rodenticides, nematicides and others make up the rest. The plantation sector especially rubber, oil palm and cocoa, is the largest user of herbicides such as paraquat, glyphosate, 2,4-D, diuron, methamidophos, picloram and dalapon.

The use of pesticides in Malaysian agriculture is regulated under the Pesticide Act of 1974, under the Pesticides Board of Malaysia, Ministry of Agriculture, and the Food Act of 1983 which regulates the amount of pesticide residue in foods. Despite a raft of regulations, however, Cheah and Lum (1993) noted a range of studies which indicated that a substantial percentage of rice farmers develop symptoms associated with pesticide poisoning. Surveys conducted in the Muda area on pesticide poisoning revealed that 51.3% of rice farmer respondents reported that they had experienced symptoms associated with pesticide poisoning. The highest incidence (24.8%) was due to herbicides (Ho, 1994; Ho *et al*, 1990).

Pesticides are also used in market gardening for the domestic or export market. In the Cameron Highlands, where temperate vegetables are grown intensively, pesticide poisoning is not uncommon. Pesticide residues were also reported on vegetables imported by Singapore from Malaysia (*New Straits Times*, 9 December 1988), which were confirmed by samples analysed by the Agricultural Department. Organophosphorous insecticide residue exceeding the minimum recommended level of 1 mg/kg were found in vegetables; some contained residues as high as 30 mg/kg. Pesticides are also used in the storage of rice grains for post-harvest protection.

Clearly, then, the widespread use of pesticides has important implications for water discharge. There is great concern about contamination of pesticides in the agricultural environment. Lindbane residues have been detected in soil, water and fish samples from rice fields in Tanjong Karang (Lee & Ong, 1983). The National Electricity Board found the presence of diazinon residues ranging from 0.00091 mg/l to 0.0037 mg/l in water samples taken from the Ringlet reservoir in the Cameron Highlands. Such values are sufficient to affect fish life in the reservoir. Low levels of profenophos residues were also detected. Mushrooms cultivated on paddy straw waste have also been found to contain organochlorine pesticide residues, indicating food chain effects.

Research has also shown that the contamination of soil, water and fish in rice ecosystems may arise as a result of pesticide use for pest control. Some of the pesticides are persistent in the environment and, like endosulphan and azinphos-ethyl, are toxic to paddy field fish (Ooi & Lo, 1990). Analysis of dead fish tissue showed high levels of endosulphan I (0.8 mg/kg) and endosulphan II (0.3 mg/kg). The danger to humans is real given that fish culture is still an important practice in the northern rice-growing areas of Peninsular Malaysia. In Malaysia, widespread use of DDT has been discontinued and is now restricted to malarial vector control programmes. Given the high rainfall and leaching and the solubility of such chemicals, much more pesticide in terms of dosage and frequency is being used by plantations and farmers, which invariably find its way into the streams and rivers and ultimately into the marine environment. Eastern Sumatra, especially in the provinces of North Sumatra and Riau, has the largest acreage of plantation agriculture, especially oil palm and rubber and as such the use of fertilizers, herbicides, pesticides, rodenticides and fungicides is widespread. Table 10.3 provides an indication of pesticide usage.

Pig rearing is also a major source of agricultural waste in Malaysia. The pollution load attributable to pig farms was estimated to be 218 tons/day or about 46% of the total load generated in the whole of Malaysia. The Linggi and Langat rivers are especially affected as many pig farms are located in their catchments. The situation will become more serious with a projected increase in the pig population in future. In 1991, some 2.3 million pigs were reared in 3,226 pig farms nationwide and the number will increase to 3 million by the year 2000.

One major concern is the number of unlicensed pig farms which have sprouted in many parts of the country. Recently, an outbreak of *Japanese encephalitis* spread by mosquitoes that breed in unhygienic pig farming environments caused several

Table 10.3 Pesticide use in agriculture in the east coast of Aceh, North Sumatra and Riau provinces (tons)

Coastal area regency	*Insecticide*	*Fungicide*	*Rodenticide*	*Herbicide*
Aceh				
Pidie	2,600,295	100,200	–	–
North Aceh	585,704	142	–	1,100
East Aceh	239	6,100	–	2,305
North Sumatra				
Langkat	129,050	632	3,329	3,770
Deli Serdang and Medan	856,461	3,183	283,813	7,712
Labuan Batu	67,502	37	3,076	4,928
Asahan	140,453	175	1,138	6,450
Riau*	85,000†	–	4,000	–

*Regencial data of Riau Province on the use of pesticides are not available.
†Including small amounts of fungicide.
Source: Chua *et al* (1997); Dahuri and Pahlevi (1994); Soegiarto (1987).

deaths. Arising from this, checks found that in Malacca alone, 102 pig farms did not have licences to operate. The same concern was expressed in Selangor and other states. Because of the pollutive nature of pig farming, Singapore has banned pig farming altogether. In Sumatra, effluent from pig farms does not pose any major problems.

Other than pig rearing, poultry and some dairy farming are the next two important activities in the areas bordering the Straits. Of the two, poultry farming for broiler and eggs is more widespread and, even in Singapore, there were some 2,000 poultry farms with a total of 3 million layers, 3.5 million broilers, 371,000 breeders and 918,000 ducks in 1989 (Ministry of Environment, Singapore, 1989). However, there has been little research on the extent of waste generation and pollution of river courses by these activities in the three countries. Certainly, as far as duck farming on the west coast of Peninsular Malaysia is concerned, much of this activity is conducted around mining pools in Perak, Selangor and Negri Sembilan and the discharge of wastes is largely confined to the pools concerned.

Pig waste together with fertilisers used in agriculture and discharge of sewage all contribute to the increased input of nutrients into the coastal waters of the Straits. Such over-enrichment of the waters through excessive nutrient input might have been the cause of extensive algal blooms in sea water, some of which are toxic to fish and other marine organisms and sometimes to human consumers (UNEP, 1982).

INDUSTRIAL WASTES

With the rapid industrialisation of the region in the last few decades a huge range of factories have been established, especially in western Malaysia and Singapore, and these have created a range of waste disposal problems which have impacted on water quality in the rivers flowing into the Straits. The need for toxic factory waste to be treated properly has prompted the Malaysian government to establish an integrated toxic waste centre in Shah Alam, Selangor. This integrated toxic waste centre incineration plant, Malaysia's first, began operation in March 1997. Four major facilities have been established:

1 A solidification plant which stabilises liquid waste or semi-liquid sludge, which is then converted into solid waste that can be safely disposed of. This plant is capable of processing 15,000 tons of waste annually.
2 A physical/chemical treatment plant which treats inorganic liquid waste, spent acid, and alkaline, cyanide and chromate waste. It can treat 5,000 tons a year.
3 An incineration plant which treats all organic waste. It is also used to destroy combustible materials such as drums. The resulting ash is packaged and disposed of. It takes much less space than the original waste.
4 A landfill capacity of 30,000 tons to contain packaged ash from the incineration plant.

The whole centre is able to treat 107 scheduled wastes listed in the Environmental Quality (Scheduled Wastes) Regulation of 1989. So far, some 400 factories have signed to have their waste treated and disposed of at the centre (*New Straits Times*, 24 June 1998). As of the end of May 1998, the landfill received 22,140 tons of various waste, the solidification plant 15,518 tons; the incineration plant 900 tons and the physical/chemical treatment plant 17 tons.

This centre marks an important first step towards tackling toxic waste disposal by industriy. But the number of factories throughout the west coast of Peninsular Malaysia is much larger than the 400 that have signed thus far. Unless toxic waste treatment is made mandatory for factories, and stringent monitoring conducted, the costs of treatment (which escalate with distance from the centre) mean that many factories will find other ways of disposing of wastes. Certainly factories located outside Selangor will find it uneconomic to tranport their wastes to the Shah Alam treatment centre. Already there have been serious cases of waste being dumped illegally. In two villages in Plentong, Johor, recently more than 200 drums containing toxic waste and metal sludge (both of which are scheduled wastes) have been dumped illegally in the village of Kampong Lunchu and Kampong Tengah. The dumpsites were located near Sungai Lunchu, a tributary of Sungei Tebrau (*New Straits Times*, 29 December 1998).

In Johor, at the southern end of the Straits, the Pasir Gudang Local Authority found that 43 of the 239 factories in the authority had been involved in water pollution incidents in 1997, largely through discharging waste into drains or rivers. The pollution was mainly caused by oil-based, coloured, acidic and alkaline effluents discharged into the waterways by the factories. Pasir Gudang is Johor's premier industrial estate, with some 239 factories, out of which 46 are chemicals, gas and fertiliser plants, 40 construction and engineering materials facilities, 26 oil and grease factories and 24 electronic plants (*New Straits Times*, 29 June 1998). The types of factory reflect the kinds of potential pollution to the Straits environment if stringent controls are not imposed.

Indonesia has taken steps to improve the management of her environment through the establishment of an environmental protection agency (BAPEDAL), environmental impact assessment requirements for development projects and new legislation in spatial planning. Through its PROKASIH (Clean Rivers) Programme 8, most industrialised provinces have initially been earmarked for this new initiative, including the province of North Sumatra bordering the Straits. Under the programme local PROKASIH teams were established, specific firms in highly-polluting industries were identified and these firms were to sign voluntary 'Letters of Commitment' to cut pollution loads in half within an agreed time frame. Subsequent progress was monitored and pressure was applied on those establishments that failed to meet the commitment. From the original eight provinces the programme now has been extended to 11, with some 2,000 firms having signed voluntary agreements. Consequently, pollution loads have been reduced in several provinces and the target was to reduce industrial pollution in 24 most polluted rivers across Indonesia (World Bank, 1994).

Heavy metals contamination

Heavy metal pollution in coastal waters is generally associated with industrial and urban areas, and the Straits waters and region are no exception. Levels of heavy metals over interim standards set were found on the coasts of Johor, Pulau Pinang and Perak in the mid-1990s, as Table 10.4 indicates. The above data comes from the regular monitoring by DOE, but some intensive local studies have also provided cause for concern. Studies of the Kelang River and its estuary have shown high contaminant levels in fish, shellfish and sediment, although these levels might have reflected antecedent conditions (Law & Singh, 1987; 1991). Researchers from the University of Technology, Malaysia have also found levels of heavy metals in shellfish which exceeded the Ministry of Health's recommended levels for food (*The Star*, 17 May 1993, 13).

Lead levels exceeding 0.01 mg/l were recorded off Perak and Penang Island and mercury levels exceeding 0.001 mg/l were frequently recorded in the coastal waters off Melaka, Selangor, Pulau Pinang and Peraks, states that abut the Straits. The coastal waters of Penang also recorded high levels of copper. The high levels of heavy metals indicate that the degree of waste treatment from industrial establishments still leaves much to be desired.

In the coastal waters of eastern Sumatra heavy metal contamination has been found in localities associated with hydrocarbon exploitation such as Lhokseumawe in North Acheh and Asahan and Deli Serdang in North Sumatra. Heavy metals have also been found in sediment samples of the Pakning River, Bengkalis, Riau Province, where major activities relating to oil exploration and production are carried out (Abdullah, *et al*, 1995).

Table 10.4 Heavy metals in the coastal waters of the Straits, Peninsular Malaysia

Metal	*State with samples above standard*	*Highest reported level*	*Lowest reported level*	*Number of samples above standard*
Arsenic (As) 0.05 mg/l	Johor (entire state)	1.83	0.037	2
	Perak	0.38	Below detection	31
Lead (Pb) 0.09 mg/l	Perak	0.33	0.14	43
Chromium (Cr) 0.1 mg/l	Perak	0.62	0.02	8
Mercury (Hg) 0.001 mg/1	Johor	2.01	0.20	23
	Pulau Pinang	0.05	Below detection	66
	Perak	0.08	Below detection	20
	Kedah	0.04	0	14
	Melaka	0.018	0.003	6
	Negri Sembilan	0.004	0	16
Cadmium (Cd) 0.005 mg/1	Pulau Pinang	0.20	0	2
	Johor	0.07	Below detection	4
	Selangor	0.27	0	1
	Kedah	0.01	Below detection	6
	Perak	0.02	Below detection	33

Source: Dow (1997, 83).

Intensive port activities and the petrochemical industry have also left their mark in terms of heavy metal pollution of the coastal waters of Singapore. Above average concentrations of copper, cadmium, cobalt, nickel, lead and zinc have been found in the waters off the south coast of Singapore (Grace *et al*, 1987). High copper and zinc levels have also been found in Keppel harbour, while high lead, nickel and cobalt were reported in waters near petroleum refineries. One comforting note is that present levels of heavy metal pollution in the coastal waters of Singapore are no worse than those found some 20 years ago, despite much increased levels of overall activity (Tang *et al*, 1996).

Air pollution

The input of substances such as heavy metals like zinc, cadmium, lead, mercury and selenium by aerosols into the open oceans can account for more than 50% of the terrestrial influx (United Nations Environmental Programme, 1982), underlining the significance of pollutant transport by air to the ocean environment. In a narrow sea like the Straits of Malacca, however, the amount and impact of atmospheric input of pollutants are difficult to determine or quantify. There can be little doubt that the frequent burning of forests and rubbish in Peninsular Malaysia and Sumatra, and the amount of carbon produced as a result that finally settles in the Straits, may be important particularly during the prolonged smoke haze episodes of 1994 and 1997–1998. The input of pollutants from vehicular combustion and factories in the Klang valley, the oil refineries in Singapore and at Banda Aceh and Dumai in Sumatra, where different kinds of chemicals are released, may cause other types of air pollution to reach the Straits.

In a climate characterised by frequent calms, the occurrence of sea and land breezes may produce some movement of air to and from land and the Straits, and thus land breezes may help transport some of these air pollutants into the Straits. Perhaps of greater significance are the volcanic eruptions which Sumatra is prone to experience; but even these are too few and far between to cause any major deposits into this water body. Thus, while theoretically it is conceivable that acid rains may have been experienced in the Straits, there is no data to substantiate its occurrence let alone quantify its intensity. Inferences from the level of air pollution in the Klang valley of Selangor, Peninsular Malaysia indicate that air pollution levels might be high in stable conditions.

SEA-BASED POLLUTION

Sea-based pollution in the Straits of Malacca comes from several sources, ranging from the infrequent but dramatic, such as tanker collisions, to the much more frequent, small-scale events such as operational discharges, deballasting, tank cleaning, bilge water and sludge discharge, and discharges from numerous small vessels. Though individually these discharges are small, as a whole the effects on this semi-enclosed sea are a cause for serious concern.

Given the heavy and increasing traffic volume through the Straits, the risk of ship collision is high. As Chapter 8 emphasised, the rapid economic development of the Straits region has greatly increased traffic densities. Serious vessel collisions that have taken place in the past include *Showa Maru* in January 1975 and the *Diego Silang* with extensive oil spillage, the *Royal Pacific* and *Tefu 51*, *Nagasaki Spirit* and *Ocean Blessing*, and *Mearsk Navigator* with *Sanko Honour*, two of which were loaded tankers and one a passenger cruise ship resulting in major oil spills and loss of lives. The last three occurred in 1992 within a few months of each other (Raja Malik, 1995).

Singapore faced the challenge of the worst recent oil spill, of 28,500 tons, as a result of a collision between an oil tanker, *Evoikos*, and an empty very large crude carrier, (*VLCC*) *Orapin Global*, in the Singapore Strait on 15 October 1997. Because of the quick response and effective cleaning measures, serious environmental consequences were avoided. More recently, on 26 September 1998, an Indian cargo ship, *ICL Vikraman*, and a Panama-registered bulk carrier, *MV Mount One*, collided off Tanjung Tuan in Port Dickson, leading to the loss of 29 lives and the sinking of the cargo ship. Apart from collision, grounding is a frequent hazard in the Straits. The incidence of haze increases the risks of collision and casualties in the Straits; in September 1997, haze problems were such that the Klang Port Authority considered closing night shipping in South Port as visibility fell below 0.5 nautical miles (*New Straits Times, Sunday Magazine*, 7 December 1997).

Vessel accidents in the Straits have been increasing as traffic densities grow, as we argued in Chapter 8. Given the narrow channel and, at some points, the shallow available water depth, large vessels are prone to grounding. In most of these cases a combination of factors come into play, apart from hydrographic conditions. Factors like visibility, currents, tides and human error are equally important. Collisions account for some 35% of all accidents in the Straits, mainly because of the high traffic density. More frequent accidents involve cargo ships and oil tankers, with loaded oil tankers accounting for 17% of ship casualties in the Straits (Lee, 1994). From 1978 to 1994, 476 maritime casualties have occurred in the Straits (Lloyds Maritime Information Service), with general cargo vessels accounting for more than half of all vessels (53%), followed by oil tankers (21%) and bulk cargo carriers (7%). In the past decade there has been an increase in accidents involving all types of vessels, especially fishing boats and small cargo ships, reflecting the increasing importance of cross-Straits movement.

As we showed earlier, the intensive use of the Straits for international and intra-Straits navigation, fishing and leisure cruises makes them one of the busiest in the world and renders them vulnerable to pollution risks from sea accidents. The volume of traffic, which is the very lifeline of the waterway, has become its own Achilles' heel, opening the Straits to the threat of real environmental degradation. Oil pollution in the Johor Straits is quite common, particularly at Johor's main port of Pasir Gudang; 48 oil spills involving foreign vessels were reported from 1979 to 1987 (International Centre for Living Aquatic Resource Management (ICLARM), 1991).

This potential threat to the environmental wellbeing of the Straits has galvanised concern from interested parties, namely littoral states, user states and international companies, as well as regional and international organisations. A considerable number of conferences[1] have been organised in Malaysia, Indonesia or Singapore to discuss these issues and to propose plans to tackle common problems.

The Straits of Malacca are essential to world shipping and to the development of

[1] The more important meetings in the past 15 years include, among others, the following: Symposium on Research and Coastal Zone Management Problems in the Straits of Malacca and Singapore, Medan, Indonesia, November 1985; SEAPOL International Conference on the Implementation of the Law of the Sea Convention in the 1990s: Marine Environmental Protection and Other Issues, Denpasar, Indonesia, 28–30 May 1990; National Conference on the Strait of Malacca, Malaysian Institute of Maritime Affairs, Kuala Lumpur, 11 November, 1993; SEAPOL Singapore Conference on Sustainable Development of Coastal and Ocean Areas in Southeast Asia: Post Rio Perspectives, Singapore, 26–28 May 1994; The First International Conference on Maritime Crisis Management: the Aftermath of Maritime Disasters, The Maritime Law Association of Singapore and the Maritime Law Association of Malaysia, Kuala Lumpur, 26–27 September 1994; Navigational Safety and Control of Pollution in the Straits of Malacca and Singapore: Modalities of International Co-operation, The Institute of Policy Studies and the International Maritime Organization, Singapore, 2–3 September 1996; Maritime Institute of Malaysia Workshop on ASEAN Maritime Security, Kuala Lumpur, 8–9 October 1996.

the littoral states and regions beyond. Shipping is of vital importance to the littoral states for all of Malaysia's main ports are situated on the Straits, as are the major Indonesian oil exporting port of Dumai and trading port of Belawan. Singapore's position as the major bunkering port in the world and a major transhipment point depends on traffic through the Straits of Malacca. Shipping through the Straits is also of vital importance to Japan, as 90% of all its oil imports pass through this narrow waterway.

A sense of the importance of the Straits to both ports and shipping can be captured from the material in Chapter 8. It is evident that the Straits area is of vital importance, not only to the countries of the region, but to the majority of the world's trading nations and it is essential that world shipping is able to navigate through this area with safety. However, given the potential for imperceptible as well as disastrous pollution of this waterway, the passage of ships may become more difficult in the years ahead. While closure of the Straits would not be acceptable to the littoral state of Singapore, because of the enormous importance of this waterway to her ports, the imposition of charges for use of the Straits may be a difficult proposition to implement and enforce as well as being unpopular to users. Any increase in costs of navigation will have repercussions on shipping in and through the Straits, and therefore economic costs to the littoral states. Reconciling the economic demands on the Straits with the environmental conditions of their waters remains a major dilemma for the three littoral states.

A range of recent regional and global incidents have made clear the disastrous consequences to marine life and beaches of major oil spills as a result of tanker collision. The impacts of oil spills may include alterations to habitat, impacts on the growth, physiology and behaviour of organisms, and mortality and morbidity of some species as a direct result of acute exposure, or as an indirect consequence of habitat alteration or loss (Dow, 1997). Oil spills have caused the death of wild fowl and marine mammals, as well as the loss of mangrove, seagrass and coral reef habitats (Goh, 1997). Much research has been undertaken to assess the short- and long-term damage on marine ecology as well as the coastal ecosystems. While some studies have shown the resilience of marine systems in being able to regenerate after a few years of such a disaster, the net effects are invariably costly. Thus, in 1987, two major spills in Indonesian waters drifted towards the beaches of southwest Johor. The spills, involving fuel oil and crude oil, adversely affected the ecology and fish catch in the area (ICLARM, 1991).

Heavy use of the Straits for navigation and by other vessels undoubtedly threatens the health of the maritime environment through ship accidents. But the environmental condition of the Straits can also be disturbed by other water-based activities, although they may be localised. These include off-shore petroleum exploration and production. As far as this activity is concerned, it is mainly confined to the off-shore parts of Aceh and Dumai in North Sumatra, where hydrocarbon resources are being produced. No production of oil or gas is carried out in the eastern parts of the Straits off the coast of Peninsular Malaysia. The pollution problems in the coastal waters of Singapore are, by and large, being controlled despite the high population density and intensive use of the coast (Chia *et al*, 1988). Off-shore mining of some metals is carried out on a small scale. Of greater significance is the mining of sand and gravel off-shore. The effects of mining of sand and gravel can be seen through the sediment plume produced in the water. Aqua-culture activities, discussed in Chapter 4, represent another potential source of marine pollution, especially since this activity is expanding very fast in the coastal waters of Peninsular Malaysia and Sumatra, as well as within the Straits of Johor and Singapore Straits.

OVERALL EFFECTS OF LAND- AND SEA-BASED POLLUTION

The multiple sources of pollutants from the littoral states of the Straits of Malacca suggest that the effects on the Straits are equally diverse, and the seriousness of the impact depends on the quantity and types of pollutants discharged into it. Broadly, the effects can be categorised into three: the effects on the health of the ecosystem and aquatic environment, the public health threat through the use of the sea, and a range of economic costs that include repair of damaged coasts, maintenance of ports, depletion of resources and future sustainability of economies and communities dependent on the sea, as well as the costs of cleaning-up operations and long-term rehabilitation of affected areas.

The churning of their waters, and the receipt of oily discharges, sometimes from tanker collision and more often from desludging and deliberate dumping of wastes, all contributes to the deterioration of the Straits environment. It has been estimated that about 1,000 to 3,000 gallons of oil residue is discharged into the sea with tank wash-water on a single voyage of a 200,000-ton tanker plying the Straits. Large concentrations of such ballast discharge have been discovered at both ends of the Straits (Valencia, 1991). Table 10.5 shows oil and grease content in the coastal waters of west Peninsular Malaysia. Apart from direct oily discharges, collisions of vessels can result in oil spillage and the detergents used in cleaning-up operations may have deleterious effects on marine life. In the event of oil spills sinking to the bottom, or spreading over a large surface area, the effect on fishery resources and their breeding grounds would be disastrous.

From the above, it is clear that there is a dilemma between keeping the Straits open as a sea-lane of enormous economic importance and maintaining their ecology as to be a source of fishery resources and healthy marine life. There is evidently also an inherent conflict between the interests of the private sector that sees the Straits in terms of their primary function of commerce, trade and shipping, and the government agencies that wish to see a healthy and clean waterway. These two objectives need not be mutually exclusive, as proper control and management should be able to balance both.

While oil spills from ship-based sources have their set of impacts on the marine life of the Straits, indiscriminate dumping and discharges of effluents from factories and industries, and wastes from residential areas into rivers constitute an equally potent

Table 10.5 Oil, grease and petroleum hydrocarbons in coastal waters off the west coast of Peninsular Malaysia

	Water concentration (μg/l)*		*No. of sampling stations*	
Location	*Oil and graese*	*Hydrocarb*	*Water*	*Year of survey*
Perlis				
Kuala Perlis	600–900	2.1–2.8	3	1995
Penang				
Batu Feringghi	600–1400	2.1–2.7	3	
Kuala Perai	400–800	1.5–1.9	3	1995
Juru	1,000–1,400	1.9–3.3	3	
Penang	–	5.0–31	9	1992
Perak				
Kuala Kurau	400–700	13–43	5	1995
Lumut	100–900	21–186	6	
Lumut	–	314–386	4	1992
Selangor				
Kuala Selangor	700–1000	16–22	4	
Port Klang	300–600	6.3–63	6	1995
Morib	400–1000	13–45	4	
Port Klang	–	18–129	9	1992
Morib	–	9–43	6	
North Sembilan				
Kuala Lukut	100–600	21–30	6	1995
Port Dickson	400–500	17–101	6	
Port Dickson	–	8–112	10	1992
Port Dickson	–	69 ± 46†	31	
Malacca				
Rembang	500–600	20–140	4	1995
Malacca	–	8–15	6	1992
Johor				
Muar	–	18–48	3	1992
Batu pahat	–	26–81	3	
Tanjung Piai	–	1555**	2	
Kukup	2,100–3,500	2,000–2,700	5	1995

*Unless otherwise stated, concentration is in ESSO Selingi Crude Equivalent.
†ESSO Tapis (filter)—a crude oil equivalent; ** Dual ng whole crude oil equivalent.
Source: Adapted from Norhayati (1997).

source of pollution in the Straits. As we showed earlier, these land-based pollutants will reach the estuaries and shores first, places the in-shore subsistence fishermen depend upon for their livelihood. Lessons from the early 1970s in Malaysia have demonstrated the damage such land-based pollution can cause. The Juru River in Seberang Perai, Penang, for example, became dead from the toxic wastes discharged by factories from the Perai Industrial Estate, located within its catchment.

Besides pollutants from factories, the use of chemical fertilisers and pesticides in paddy planting has not only affected the fresh water fish life in the paddy fields, but also fish production along the shore. In Penang, a study on the concentration of *coliform* and *E. coliform* bacteria, associated with human and animal wastes, found that the microbial levels in the waters off the northern coast of Penang where concentrations of fishery resources are located were extremely high (Tan, 1992). The consumption of such fishes from these waters may pose a potential health hazard to local communities.

Apart from the consequences of land- and sea-based pollution that affect the ecological health of the fishing grounds and ultimately fish catch, the seas of the region have experienced the red tide phenomenon. Past incidents of this phenomenon have been recorded in the South China Sea off the coasts of Sabah and Brunei, although much less in the Straits. The causes of this highly toxic algal bloom are still debated. There seems to be a coincidence of the bloom with rich organic discharge from estuaries after heavy rainfalls. However, not all heavy rainfalls causing heavy discharges into the sea bring about red tide phenomena. The problem with red tides is that fish and other marine life caught in such areas contain these toxic substances which can be fatal for humans who consume them.

Chia (1995) has discussed the various effects of oil spill on the marine environment in ASEAN. These include ecological effects in terms of habitat quality deterioration, leading to changes in the organisms and communities; effects on mammals, birds and fish in shallow water and on beaches as well as in mud flats and mangroves; ecotoxicological effects; bioaccumulation and tainting of seafood; and effects on humans using the coastal areas for different purposes. It is clear that oil and other pollutants will pose serious threats to all forms of life in the sea including sea grasses, plankton species, molluscs, larger species of fish, sea birds and sea turtles.

STRAITS MANAGEMENT AND POLLUTION CONTROL

The health of the Straits environment depends very much on land- and sea-based pollution generation and control. Of the two, land-based control should be theoretically easier to implement as each state has complete jurisdiction of its own territory, and the management and control of pollution can be easily regulated through its own set of environmental laws and system of fines and penalties. Of course, this is easier said than done. The lack of success of pollution control on land is very often due to the lack of political will and to deficiencies in the implementation of the laws and regulations which are already in place. However, where sea-based pollution is concerned, it is more difficult to detect and control partly because there are many externalities involved, such as ships belonging to foreign companies and discharge of pollution in international waters. Unless spillage is on a large scale which, in most cases, would be classified as a disaster, any small-scale discharge often goes unnoticed.

However, the complexity of governing and managing an international waterway like the Straits of Malacca means that all parties must be involved, and control and management must be organised at different levels and scales. In this respect, various measures must be taken simultaneously in the management of the Straits marine environment.

Given the difficulty of navigation of vessels in narrow and hazardous straits, navigational aids are an important ingredient in ensuring the safe passage of ships, particularly the larger ones. Such aids have been set in place over the years and improvement in equipment is constantly being made. Thus far, the majority of aids to navigation (buoys/beacons) in the Straits of Malacca have been funded by the Malacca Straits Council of Japan which, as we noted in Chapter 8, had a comprehensive reinstallation schedule covering planned work on navigation beacons up to 1995. The International Maritime Organisation (IMO) Working Group on the Malacca Straits visited Malaysia in 1993, and was informed that Malaysia is maintaining 99% availability of off-shore aids to navigation and this will be upgraded to 99.9% availability and reliability. Indonesia maintained that there had been no complaints regarding the failure of aids to navigation in the TSS (Traffic Separation Scheme) area. One major problem connected with navigational aids is that unidentified vessels may collide with navigational buoys causing them to be damaged or dragged out of position. In 1992, a safety study of navigational risks in the Malacca/Singapore Straits by the Oil Companies International Marine Forum (OCIMF) concluded that a significant safety improvement would result if selected navigational buoys were replaced with fixed light beacons or fixed radar transponder beacons.

Navigation aids assist vessels to recognise designated routes, but negotiating through the specified routes must also be governed by the observation by vessels of certain rules. The Japanese have stressed that navigation safety is the foremost common interest among all the littoral and user states. Two ways of going about achieving this objective as far as the Straits of Malacca are concerned are: (1) to improve the conditions of the Straits, establish routing and traffic separation schemes, deep water routes and special rules and, (2) to require under-keel clearance of fixed metres, thereby forcing very large crude carriers (VLCCs) to take safer but longer alternative routes such as the deep water corridors of the Lombok and Makassar Straits.

Following the *Showa Maru* grounding in 1975, the three coastal states agreed to introduce a traffic separation scheme (TSS). This scheme was further improved in the 27 February 1997 agreement on safety of navigation in the Straits of Malacca and Singapore signed by the foreign ministers of the three countries. Two recommendations were carried through: that, 'vessels maintain a single Under Keel Clearance (UKC) of at least 3.5 metres at all times during the entire passage through the Straits of Malacca and Singapore' and that a TSS incorporating two deep water channels be delineated in three critical areas – One Fathom Bank (off Port Klang), the main Singapore Strait and Phillip Channel at its western entrance, and off Horsbourgh Lighthouse near the entrance to the South China Sea. For the TSS, vessels with a draft of 15 m and above were to use the designated deep water route, and to travel at not more than 12 knots during their transit through the critical areas.

There are areas outside these schemes where navigation for deep-draught vessels is particularly difficult and further assistance to shipmasters is recommended to enhance safety of navigation and protection of the marine environment. Such areas are off Medang and Cape Richardo, where large vessels are required to make major course alterations in relatively narrow waters. The OCIMF/ICS Publications (1990) advises a recommended track 'off Tanjung Medang' which is of value to shipmasters. As we noted in Chapter 8, it seems likely that future negotiations between the states and users will result in the designation of additional traffic schemes in the Straits. Following the decision of the 18th Tripartite Technical Experts Group (TTEG) meeting held in 1993 in Kuala Lumpur, a working Group of the TTEG was formed. The group met in 1994 in Langkawi to consider the proposal to review the existing routing system. A proposal was made to the three governments which endorsed the recommendations of 1994, and subsequently Malaysia submitted a proposal for new routing measures in the Straits of Malacca to IMO's Sub-Committee on Safety of Navigation at its 41st session in September 1995.

Major benefits would accrue if ships plying the Straits could be in radio communication with littoral states at all times. At present no such vessel traffic system covering the whole route is in place. Only parts of the Straits of Malacca are covered by the Singapore Vessel Traffic Information System (VTIS) and by the Indonesian Customs Coastal Radar System. The comprehensive radar and computer-based VTIS covering the Singapore Strait Traffic Separation Scheme has been in operation since October 1990, and although only 20 to 30% of through traffic participates in the VTS, it is a useful mechanism for improving safety of navigation in the Singapore Strait. The use of the Differential Global Positioning System (DGPS) to provide more accurate positioning data for ships plying the region was introduced in Singapore in October 1998. This free service is offered to shipping companies and the public on a round-the-clock basis.

In Indonesia, a shore-based high resolution radar system was established mainly for anti-smuggling purposes. Radar scanners are located on the highest points of the Karimun Island and Batam Island and provide a coverage of 32 nautical miles. Because of the specific use for which the system is intended, communication is confined to customs patrol boats only; communications with agencies that could contribute towards the monitoring of vessel traffic in the southernmost parts of the Straits is non-existent. Malaysia may introduce a sea surveillance system initially involving radar coverage at One Fathom Bank, south of Kelang, Port Dickson and Malacca.

Chia (1997) strongly advocated the implementation of the Electronic Chart Display and Information System (ECDIS) which gives real-time locations of ships. However, with improved technology a more comprehensive information system which integrates telecommunication technologies, ECDIS and Electronic Navigation Charts (ENCs), as well as the transmission of real-time water level and current information could be implemented under the concept of a Marine Electronic Highway

(Macdonald & Anderson, 1996). This system would have significant anti-grounding and anti-collision capabilities.

The dynamic nature of the sea bed and shoal patches in the Straits makes it imperative that regular surveys are conducted to determine their shifting positions and hence available water depth. The last surveys of the Straits were conducted between 1968 and 1978 by the Malacca Straits Council together with the three littoral states, and Common Datum Charts were produced. There is now an urgent need to re-survey certain areas in the Malacca and Singapore Straits as the prevailing situation is exacerbated by the presence of large patches of dynamic sand waves making the depth rather unpredictable. Some areas are critical, with a potential for accidents, such as between One Fathom Bank and Tanjung Piai—three collisions in the area in 1992 resulted in loss of lives and severe marine pollution.

The use of satellite remote sensing is becoming an effective tool in identifying and tracking pollution from both land and sea. Malaysia is applying this technique to monitor illegal clearing of forests, the progress of large-scale developments on land and tracking of pollution in the Straits. Singapore, through the joint Memorandum of Understanding between the Maritime and Port Authority of Singapore and the Centre for Remote Imaging, Sensing and Processing (CRISP), will use satellite imagery for observing ship movement, port congestion and searching for missing vessels. This technique can also reveal oil slicks on the sea surface, thus deterring illegal discharge of oil by vessels. This will enhance current efforts by the Maritime Port Authority of Singapore patrols to combat pollution in Singapore waters.

All the measures taken above, from providing navigational aids, routing, hydrographic surveys to providing information of ships and to ships, and surveillance, cost large amounts of investment not only in terms of initial costs but on-going ones as well. To illustrate this point, Malaysia's expenditure on the above measures is reflected in Table 10.6. The above costs do not include what Indonesia and Singapore spend in their parts of the Straits.

Navigational safety and pollution control are intertwined and to maintain both exerts a heavy burden on littoral states. Safety becomes a paramount consideration where lives are concerned and more and more people are using the Straits for various purposes. Apart from sailors and crew of ships and tankers, or the fishermen, increasing affluence has brought about rapid demands and growth of the marine recreational and leisure industry in Malaysia and Singapore. Safety must also take into consideration the navigational safety of cruise liners that increasingly make their presence felt in the Straits, as well as of individuals using jet-skis and personal watercrafts by the coast (Wong, 1996). When there are major oil spills the costs will escalate in clean-up operations alone. Table 10.7 shows the estimated costs incurred by Malaysia in the clean-up operations of some major oil spills. Costs will also be incurred by the littoral states long after the clean-up operations are completed. It is not surprising, then, that Malaysia has floated the idea that a levy should be imposed on every vessel that plies the Straits so that this money could be used for clean-up operations as well as to compensate littoral states for damage to the marine and coastal environment.

Table 10.6 Safety measures and their costs in the Straits of Malacca, Malaysia

Navigational aids

Type	*Cost/unit (RM)*	*Number of Units*	*Total cost (RM)*
Lighthouses	2,400,000	10	24,000,000
Beacons	100,000	103	10,300,000
Buoys	45,000	143	6,430,000
Racons	75,500	3	225,000

Surveillance (1989–1993)

Year	*Cost (RM)*
1989	7,810,000
1990	8,220,000
1991	8,650,000
1992	9,110,000
1993	9,590,000

Marine search and rescue operations (1989–1993)

Year	*Number of incidents*	*Cost (RM)*
1989	52	731,952
1990	61	858,636
1991	66	929,016
1992	86	1,210,536
1993	82	1,154,232
Total	347	4,111,602

Survey costs (1984–1992)

Year	*Cost (RM)*
1984	5,660,000
1985	7,350,000
1986	7,950,000
1987	6,750,000
1988	6,350,000
1989	6,950,000
1990	7,550,000
1991	6,050,000
1992	7,050,000

Source: Muhammad Razif Ahmad (1997).

The need for the international community and the user states and shipping companies to work with littoral states to ensure that the use of the Straits is not disrupted cannot be overemphasised. While the littoral states should be primarily responsible for legislation and policies to curb pollution from land-based sources, managing pollution from vessel sources demands cooperation from all parties at the national, regional and international level (Hamzah Ahmad, 1997).

Table 10.7 Clean-up costs of major spills in the Malaysian waters of the Straits

Clean-up cost of major oil spills

Name of ship/flag/year	*Amount of spill (Metric tons)*	*Clean-up costs (RM)*
Showa Mary/Japan/1975*	17,700	1,250,000
Diego Silang/Philippines/1976	5,500	2,661,731
Nagasaki Spirit/Panama/1993	13,000	3,690,093
Others (minor)	13,800	113,436
Total	50,000	7,715,260

Annual maintenance cost (1989–1993) in RM

Year	*Oil Spills*	*SAR*	*Surveillance*	*Surveys*	*Navigational aids*	*Total*
1989	0.38	0.73	7.81	6.95	4.91	20.78
1990	0.15	0.86	8.22	7.55	4.76	21.54
1991	0.23	0.93	8.65	6.05	7.35	23.21
1992	3.99	1.21	9.11	7.05	7.22	28.58
1993	0.75	1.15	9.59	7.55	7.83	26.87

*Total damage claimed by Malaysia amounted to over RM25 million.
Source: Muhammad Razif Ahmad (1997).

INTERNATIONAL CONVENTIONS, REGIONAL AND INTERNATIONAL COOPERATION

Because of the complex nature of the issues connected with oceans and sea management, an international approach to managing the marine environment in terms of access, resource exploitation, waste and pollution control and piracy is vitally important. As far as environmental issues are concerned, there are several conventions dealing with pollution of the seas, starting with the significant International Convention for the Prevention of Pollution from Ships in 1973, as modified by the 1978 Protocol (MARPOL 73/78). But prior to this, two Law of the Sea (LOS) conferences were held under the auspices of the United Nations. The first was held in 1958, which drafted four conventions dealing with the high seas, the territorial sea and contiguous zone, fisheries, and the continental shelf, with no concensus on the width of the territorial sea. The second conference was held two years later, which also failed to resolve the territorial sea issue. The third and most significant was the United Nations Conference on the Law of the Sea (UNCLOS 3) in 1973.

As we indicated in Chapter 9, it took nine more years before an agreement was reached on a new 'constitution' for the world's oceans, which defined a 200-nautical mile Exclusive Economic Zone (EEZ), whilst guaranteeing most navigational freedoms, and a 12-nautical mile maximum width of territorial seas. The convention finally came into force on 16 November 1994, after the 1992 United Nations Conference on Environment and Development (UNCED) in Rio provided more detailed guidance regarding options and management approaches for national ocean zones (Cicin-Sain & Knecht, 1998). As far as the Straits of Malacca were concerned, the relevant provisions of UNCLOS include greater security for rights of passage, protection of the marine environment and mandatory settlement of disputes. As there are now clear sets of legal rules governing the use of the Straits, which set out the rights, jurisdiction and obligations of states, there is now a greater potential for closer cooperation between the littoral states and any misunderstandings can be resolved by peaceful means. It is in the interest of all the three littoral states of the Straits that this should be so.

Outside the more specific conferences noted above, maritime issues of pollution were discussed under the broad issue of environmental problems which culminated in the 1972 Stockholm Conference or the United Nations Conference on the Human Environment. The conference helped to accelerate the adoption of several international agreements dealing with ocean dumping and vessel-source pollution among others. The fact that ships were recognised as sources of pollution was significant, as was the recognition of the need for new regulations to deal with the issue. In the same year (1972), at the London Convention the first global standards governing the dumping of wastes into the oceans were adopted.

The United Nations Conference on Environment and Development (UNCED) in 1992, better known as the Rio Summit, was in part a restating of the issues discussed some 20 years earlier in Stockholm. Out of the many conventions on a wide range of environmental and development issues came the Agenda 21 Action Plan for Environment and Development Issues, signed by 172 nations and it is in Chapter 17 (Protection of the Oceans, All Kinds of Seas, Including Enclosed and Semi-Enclosed Seas, and Coastal Areas and the Protection, Rational Use and Development of Their Living Resources) of the 800-page document that the interests of the oceans are captured. This chapter spells out seven major programme areas:

1 Integrated management and sustainable development of coastal areas, including Exclusive Economic Zones.
2 Marine environmental protection.
3 Sustainable use and conservation of living marine resources of the high seas.
4 Sustainable use and conservation of living marine resources under national jurisdiction.
5 Addressing of critical uncertainties in management of the marine environment and climate change.
6 Strengthening of international cooperation, including regional cooperation and coordination.
7 Sustainable development of small islands.

In essence the text stresses the need at all levels for integrated, pro-active approaches to marine and coastal resources management (CRM), and provides a series of suggested actions that can assist coastal nations in strengthening their CRM efforts. No doubt these issues are likely to be of major relevance to the Straits of Malacca in future years.

Cooperation between the three littoral states of the Straits of Malacca in the control and prevention of pollution within this common waterway began in the early 1970s, although legal issues of territory and jurisdiction continue to hamper full cooperation. However, increasingly closer cooperation and collaboration have been features of Straits management and the collaboration has not been confined to this group of countries. Close cooperation in dealing with pollution levels at the sub-regional level began in 1971, when the Tripartite Technical Experts Group (TTEG) involving Indonesia, Malaysia and Singapore was established to handle all technical matters pertaining to the prevention of pollution in the Straits of Malacca and Singapore. The TTEG was instrumental in the successful adoption and implementation of the TSS in the two straits (Chua, 1995).

The fact that the three littoral states are members of ASEAN provides another reason for greater collaboration in the prevention of pollution in the regional seas, including the Straits of Malacca, not to mention the useful mechanism within the regional organisation for the resolution of disputes. Thus, within ASEAN several projects have been developed and implemented under the Committee on Science and Technology (COST), with its dialogue partners from the United States, Australia, Canada, the European Community and Japan. In 1978, the ASEAN Expert Group on the Environment (AEGE) initiated a project dealing with the technology transfer in the treatment of effluent from rubber and palm oil industries and the preparation of an ASEAN oil spill contingency plan. Another platform for collaboration is the Coordinating Body of the Seas of East Asia (COBSEA), whose many activities include oil spill contingency planning, assessment of mangrove ecosystems, EIA, and river inputs and other land-based sources of pollution.

Through funds provided by UNDP (United Nations Development Programme), a Regional Programme for the Prevention and Management of Marine Pollution in the East Asian Seas became operational in 1994. A substantive component of the programme is devoted to developing demonstrable mechanisms for the effective prevention, control and mitigation of marine pollution from land-based sources, as well as pollution risk management in international waters. A major effort is being made to reduce pollution risk in the Straits of Malacca (Chua, 1995).

Long-term collaborative efforts with Canada, Australia and the United States (especially through the ASEAN/US Coastal Resources Management Project (CRMP)) started in 1986 and had completed in 1992 six pilot sites in six ASEAN countries. These projects, though based on specific sites, provided more than just detailed base-line information about both the land and sea parts of the coastal zone of those areas. A whole range of issues including guidance in national development policy, networking among institutions in a multidisciplinary manner, linkages and coordination among government agencies and the development of a typology on Integrated Coastal Zone Management (ICZM) for future initiatives were dealt with.

One important aspect of this collaboration is the ability to respond to oil spill disasters. It is obvious that not only must all parties be involved in tackling such an eventuality, but the ability to respond quickly to deal with such emergencies is of paramount importance. Thus a system has been put in place under the ASEAN Oil

Spill Preparedness and Response Plan (OSPAR) in the wake of the spill off Indonesia by *Maersk Navigator.* In 1993, Japan provided funds to purchase oil spill control equipment.

Collaboration in pollution prevention is not confined to states or international organisations but must involve the private and shipping sectors as well. Since oil companies have a major stake in the wellbeing of the Straits, they too must be involved in developing oil spill response mechanisms at national and regional levels. A three-tier response system to oil spills has been developed where the bigger companies have their own response strategies, supported with the necessary equipment and facilities to tackle spills as a result of their own operations. Within each country, private oil companies have their own action groups such as the Petroleum Industries Malaysia Mutual Aid Group (PIMMAG), and any response to oil spills is based on pooled resources of member companies. Another example is the private, non-profit group, the East Asia Response Ltd (EARL); formed in 1992, based in Singapore, which seeks to provide an efficient and prompt response to oil spill incidents in the Asia–Pacific region. Equipment stockpiles kept by this company are able to cope with a wide range of oil spill situations and environmental conditions (Chua, 1995).

As far as the shipping sector is concerned, the Malacca Straits Council (MSC) is most prominent. Shipping companies with the support of the Japanese government have undertaken some projects directed towards the prevention and mitigation of oil pollution in the Straits of Malacca. Established in 1969 as a private corporation with the participation of the Japan Shipping Industry Foundation, shipowners and other interested parties, the MSC's purpose is essentially to ensure safety of navigation through the straits of Malacca and Singapore. As improvement of navigation routes would ensure this end, the MSC has sponsored several projects that provide better information on the hydrographic and oceanographic conditions of straits, including improving and maintaining navigational aids (see Table 9.1).

There have been calls recently by Japan for other flagship states to share in the costs of maintaining and improving navigational aids in the Straits. Japan has indicated that countries other than Japan whose ships transit the Straits of Malacca should take some responsibility for navigational safety and oil spill response services. The long-standing assistance by Japan through the MSC towards providing navigational aids in the Straits for the past 30 years has cost her some US$120 million. The Nippon Foundation has funded the building and upkeep of 41 navigational aids in the Straits. Japan remains heavily dependent on imports of many basic commodities by sea. In recognition of the large number of tankers transiting the Straits on their way to Japan, the country has contributed to the building up of oil spill response capabilities in the region, including some US$60 million worth of equipment in Singapore alone (*Business Times*, 28 January 1999).

The TSS which is supported by the Malacca Straits Council and received the approval from the IMO Assembly provides guidance to the use of the Straits of Malacca and imposes insurance and compensation schemes for vessels entering them. The council also established a revolving fund in 1981 for oil spill response which can be used by the concerned coastal states to deploy the necessary manpower and equipment to deal with oil spill. The fund is being managed in turn by the three countries bordering the Straits of Malacca.

The reduction of pollution risk at sea, particularly across political boundaries, will require the combined efforts of both the concerned governments and major users to comply with the international conventions on marine pollution, as well as to develop the necessary preventive and mitigating measures, and the financial and human resources to implement them. Sufficient provisions are already in place to take care of trans-frontier pollution of the sea, as embodied in the United Nations Conference on Human Environment (1972, Stockholm) and the Law of the Sea Convention in 1982. Clearly, there is sufficient evidence to establish the existence of a rule that prohibits the injurious use of territory 'in such a manner as to cause harm to the territory of another state' under general international law and this rule is applicable to all types of trans-frontier pollution, regardless of whether the pollution occurs in fresh water, in the air or in the marine environment (Meng, 1987).

Despite this legislative framework, an effective operational model has yet to be developed, although such a model is slowly emerging in the case of the Straits of Malacca. This would include:

- Development of oil spill contingency plans at national and regional levels.
- Establishment of the three-tier response system by the private sector.
- Establishment of national and regional stockpiles of oil spill response equipment.
- Establishment of the oil spill revolving fund.
- Availability of international conventions and protocols on marine pollution.
- A viable traffic separation scheme.
- Strengthening of national and regional technical capability (Chua, 1995).

Singapore has become a party to the 1992 Protocol to the International Convention on Civil Liability for Oil Pollution Damage, originally passed in 1969. The Instrument of Accession to the Protocol was deposited with the International Maritime Organisation (IMO), which is the depository for the Protocol, on 18 September 1997. The Protocol came into force for Singapore on 18 September 1998. What this protocol means is that shipowners are required to take out liability insurance for pollution damage caused by oil spills. Singapore-registered oil tankers are covered by insurance for a higher limit of compensation and foreign tankers calling at Singapore are properly insured for a higher limit against oil pollution damage in Singapore waters.

SEA-LEVEL RISE

Sea-level rise as a result of global warming is of concern to the coastal areas bounding the Straits of Malacca. Although the prospect of sea-level rise in the next few decades seems certain, the extent and regional impact of such a rise is poorly understood.

One major concern is with the response of mangroves to rising sea-levels. As we noted in Chapter 3, the mangrove forest on the west coast of Peninsular Malaysia has been contracting in face of demands for development of settlements and ports. Apart from this source of pressure, erosion is threatening more than half of the mangrove coast of the peninsula. The states experiencing the most extensive erosion are Perak and Selangor. Of the eroding mangrove fringes, 53% are retreating at up to 8m/year, 11% are retreating at >8 m/year and the remaining 36% are receding but at an unknown rate (Teh & Lim, 1993, 61). Erosion of mangroves is widespread along the west coast of Peninsular Malaysia. The most severe retreat occurred along the west coast of Penang, where recessions of up to 100 m/year have been recorded. There are also advancing mangroves covering a length of some 241 km and accreting at different rates, but 38.3% per cent at >15 m/year. Some of these rapidly accreting areas are found in Kala Sanglang, Perlis (>60 m/year), Sungai Bakau, Sungai Kelumpang, Sungai Perak and Bagan Sungai Tiang in Perak, B. Beting Kepah, Sungei Tunggol, and Sungei Terap in Selangor, and Parit Burong in Malacca.

In the event of sea-level rise, most of the mangroves on the west coast of Peninsular Malaysia and along the Sumatran coast will undergo retreat along the outer margin without a corresponding retreat along its inner margin, because of the presence of bunds constructed to make possible the use of coastal land for agriculture. The ultimate effect would be the disappearance of the mangrove belt if sea-level rise persists.

It is likely that sea-level rise, coupled with mangrove retreat, would lead to the increased flooding of coastal towns, settlements and agricultural lands, particularly during high tides. Already some parts of the city of Georgetown in Penang experience inundation at neap tides. In addition, sea water intrusion would threaten the shallow ground water on which most coastal villages in Sumatra and in the Riau archipelago are dependent. Projections of the extent of sea-level rise in the next half century vary, but even a rise of 50 cm by 2025 would have a tremendous impact on the sustainability of development in Indonesia. Apart from the inundation of coastal cities (including parts of Jakarta), certain transmigrant settlements could be affected (Soemartowo, 1991) and many of these settlements are found in eastern Sumatra.

The Straits of Malacca will continue to play an important role as a major sea route for a long time, even though the inherent physical constraints will increasingly test that role in future. Fears of pollution, resource depletion and marine ecological deterioration will increase as usage of the waterway intensifies. Despite all the efforts being made by littoral states individually or collectively, or in collaboration with user or flagship states, and international organisations, the Straits of Malacca will continue to be like a sink into which all kinds of pollutants are added from both land and sea sources. As this chapter has shown, considerable efforts have been made to curb this pollution, both directly, by focusing on sources, and indirectly, by making passage through the Straits safer.

Littoral states have taken steps over the years to address this problem of controlling pollution and managing the Straits environment within their borders. Individual countries have enacted a whole gamut of environmental laws designed to reduce environmental pollution of all kinds including that of the marine environment. Singapore has been relatively successful in tackling environmental problems within its borders. The Clean Rivers Programme from 1977 to 1987 was a model of successsful rehabilitation of what was essentially very polluted river systems of the Kallang and Singapore Rivers, resulting in the restoration of biodiversity of the ecosystem. A range of actions were taken in connection with this cleaning-up of rivers. These actions included zonation, relocation of people and businesses to proper low-cost housing units, industrial estates, sewerage cleaning and rehabilitation, clearing waterways of obstruction by dredging and removal of sunken derelicts, landscaping, water quality monitoring, designing a system of fines and penalties, integrating an environment curriculum in schools, increasing public awareness, promoting private sector involvement, and compliance monitoring. After restoration, it was found that with improvement in the physical and biological conditions the river was reported to support over 20 species of soft-bottom invertebrates (Yip *et al*, 1987). In many ways, its small size enables Singapore to develop a good model for the development of the whole island state, which in many ways is synonymous with well-planned integrated coastal zone management, though whether that example is appropriate for neighbouring states is doubtful.

In Malaysia, despite significant successes in improving the quality of the environment since the promulgation of the Environmental Quality Act (1974), fears associated with environmental degradation of land and water remain as real today as they were more than two decades ago (Sham, 1997). The situation in Indonesia remains a matter of concern given its much larger size and the more daunting development problems it faces.

Despite the above assessment of efforts made by littoral states to curb land-based pollution, efforts have been made to restore some parts of the marine environment in the face of development and degradation. The merging of the four southern islands of Singapore through reclamation made it necessary for corals and other marine organisms to be transferred from these islands to another area in the hope of creating five new patch reefs (Newman, 1992). In Malaysia, the Pulau Payar Marine Park was established in 1987 within the Langkawi group of islands and this park has been attracting an increasing number of people which is a cause for concern. Subsequently the idea of a Special Area Management Plan (SAMP) covering a bigger area incorporating the coastal zone of north and south Kuala Kedah, Langkawi, the southernmost portion of Perlis and out to the marine waters and islands of the Pulau Payar Marine Park was adopted by the Department of Fisheries, Malaysia. This integrated approach takes into consideration the unique features of the area's ecosystem, as well as the people living in the area who depend on the health of the ecosystem in one way or another (Nickerson *et al*, 1997).

Both academic and governmental research strongly suggest that the range of coastal pollution problems should be addressed within the context of integrated coastal zone management (ICZM). This concept has gained some currency within Southeast Asia with six pilot projects in ASEAN countries being supported by the ASEAN/US Coastal Resources Management Project (CRMP) from 1986 to 1992. While the pilot projects provided useful pointers for future development of the coastal zone in ASEAN, it remains to be seen how many of the recommendations and benefits can be translated into standard practice for the development of coastal zones bordering the Straits. Nevertheless, a holistic, comprehensive and integrated approach to the development of coastal zones must necessarily include efforts to plan and manage human activities in the coastal areas and the wise use of the coastal and marine resources (Chua, 1995).

It is evident that the health of the Straits of Malacca should not be solely the concern of the littoral states, although it is in the direct and long-term interests of the littoral states that they play the pivotal role. Certainly a clean and well-managed coastal environment will be an asset and an attraction that brings, and will continue to bring, economic benefits to the states concerned. However, the littoral states must act with international support. Regional and international and other stakeholder involvement is vital to the successful attainment of the objectives of Agenda 21 of the Rio Summit. There is much more to be done and the tasks ahead will not be easy as the diversity of stakeholders makes the issue of funding for environmental management and navigational safety contentious (Hamzah, 1996, 7). Going by the efforts and progress made thus far, we are hopeful that future concerted efforts involving all parties will bear tangible fruit.

11

CONCLUSION

We have argued in this book that the regions abutting the Straits of Malacca, though politically and administratively divided, share a wide range of historical and contemporary characteristics, and that their landscapes, societies and economies have been shaped by the common experience of sharing that narrow, busy stretch of water. In short, a common heritage and set of opportunities outweigh the differences resulting from divergent political or economic systems. More than ever, a shared interest in the nature, use and management of the Straits will continue to characterise the regions and countries abutting these waters.

The physiographic characteristics of the region have been shaped to a considerable extent by the relationship between sea-level and the land mass. As Chapter 2 showed, the region is characterised by complex plate movements which have helped to govern the distribution and character of upland areas, as well as exerting an influence on the nature and distribution of tectonic activity, especially on the Sumatran side. Fluctuations in sea-level, especially since the Quaternary, have had an enormous influence on both the physiography and human use of the region. It has been sea-level change which has been an important control on the extent and character of mangrove forests along the coasts of the Straits, as well as on the nature of the rivers flowing into the waters themselves. The character of the sediments of the Straits, an important control on human use of the marine and coast resources, has also reflected the unifying impacts of sea-level change. Raised beaches, a general absence of cliffs and magnificent mangrove formations (especially on the Sumatran side) have had important consequences for human use of the Straits for both shipping and settlement.

The influence of climate has also shaped the unity of this region. Without being deterministic, monsoon wind patterns undoubtedly facilitated the role of the ports of the region as 'lands below the winds', transitional ports where ships could shelter whilst waiting for the winds to turn to make their onward journeys to either India and the Middle East or China. That legacy from the era of sail ships continues despite technical advances in shipping technology. Climate, too, especially rainfall, has had important common impacts on rates of weathering, runoff, deposition and soil formation.

The unifying role of the Straits was also emphasised in our discussion of landscapes, vegetation and soils in Chapter 3. The major rivers drain into the Straits which constitutes a basin-like feature drawing waters from the Main Range in Peninsular Malaysia and the Barisan Range of Sumatra. The widespread extent of mangrove and peat swamp forest in the Straits region has also resulted in common approaches and common problems in the development of peatland and mangrove areas.

If the region shares some distinctive landscape, marine and vegetational characteristics, the purpose of Part 2, *Resources and techniques*, was to emphasise that the identification and use of the range of resources in the region has been highly diverse. A common physical environment has not necessarily resulted in common patterns of use or, indeed, misuse of resources. As the historical and contemporary patterns of exploitation of, for example, the tin or the fishing resources of the region showed, resource endowments depend on human ingenuity for their exploitation and, sadly, on human greed for their destruction. Changes in technology, from the advent of new sailing ships in the fifteenth century to the arrival of steam ships, hydrocarbons or containerisation in the twentieth, have been an important agent of transformation in the Straits region, differentiating one part of the coast or interior from another, one port city advancing whilst others fall into temporary or permanent decline. From Malacca in the fifteenth century to Singapore in the late twentieth, the highly developed port cities of the region have been shaped by the constant interplay of resources, technology and human skill.

The extent to which the region has shared a common historical experience was the subject of our Part 3, *Collective histories*. In that section we sought to demonstrate that the flows along the Straits of diverse products, ideas, currencies and peoples created social, economic and political systems which were inherently diverse in their structures and open to new ideas and new economic and social currents. This is not to suggest that a single uniform process of historical change characterised the region—different resources, human endowments and socio-economic systems inevitably created diversity; but it is to recognise a number of common elements as shaping the way the societies of the region evolved. Strong maritime traditions and skills, a reliance on both short- and long-distance trade, the key geo-strategic importance of the region in relation to world trading routes, exposure to conflict, conquest and colonisation were some of the common features of the region. Our discussion of just some of the key phases in the region's history—the rise of Malacca and Aceh, the impact of the Dutch and British, the rise of Singapore as a global port—has hopefully demonstrated the advantages of viewing the region as a single unit rather than a series of discrete, unconnected areas.

The final part, *Collective opportunities*, sought to underline the common interests that the states abutting the Straits increasingly share. The rapid economic growth of Malaysia and Singapore in the last three decades, a growth built on rapid export-orientated industrialisation, has posed enormous challenges for planners and politicans alike. The high growth rates and rapid industrial and urban growth described in Chapter 8 have brought enormous benefits to many of the peoples of the region. It is also evident that, in the contemporary period, the gap between the western and eastern seaboards of the Straits has inexorably widened. A continued reliance on primary products, coupled with poorer infrastructure and development prospects have hampered growth in the east coast provinces of Sumatra. As Chapter 10 demonstrated, the environmental costs of economic growth have been high. Land- and sea-based pollution into the Straits, the problems of managing some of the world's busiest shipping lanes and important issues of agricultural and tourist development in the coastal zone are problems that increasingly require a high degree of international cooperation.

THE FUTURE OF THE STRAITS

One central conclusion of this book is the tremendous importance of the Straits as a corridor of historical development, a vital routeway for global shipping and an important factor in the economic development of the region. There can be little doubt that the economic and political turmoil of the mid-1990s, a turmoil which has embraced currency crises in both Malaysia and Indonesia and political instability in Indonesia, has created a range of difficulties for the region. But such instability has long been inherent in the region—a review of the historical geography of the region has shown how the constant rise and fall of kingdoms was an integral part of its history.

Recent turmoil though, in threatening to slow down the growth rates of the region, still needs to be put into perspective. Growth rates in Malaysia and Singapore remain well above global averages and there is likely to be only a brief lull in the pace of infrastructural change in the coastal states. For Sumatra, the picture is rather gloomier and the economic crisis in Indonesia is likely to result in a further widening of the economic gap between western and eastern seaboards of the Straits.

Perhaps one of the most pressing issues likely to face the states of the Straits region is the devising of appropriate mechanisms to manage some of the environmental issues in the waters and coastal zones. The ability of Indonesia in particular to devote the time and money to questions of ocean governance, pollution control and environmental management has undoubtedly been compromised by economic and political uncertainties. On the part of both Malaysia and Singapore, there is a determination to ensure adequate management of the coasts and waters of the Straits. As Chapter 10 emphasised, both Malaysia and Singapore have initiated a range of institutions and programmes to improve the management of the Straits environment. This issue is likely to become more, rather than less, pressing. For the foreseeable future, the Straits will remain a key artery of global shipping, especially for Japan and east Asia, and the pressures of increased shipping will require international management. The involvement of Japan in a number of important maritime safety issues in the Straits should presage the greater involvement of other major users of the Straits. Devising appropriate fora for the development of such cooperative programmes will be one of the major challenges facing the region in the future.

For well over 1,000 years, the Straits region has been home to some of the richest and most dynamic societies in the world. Animated by the currents of long-distance trade, their economic, social, political and religious structures have endowed the region with a rich heritage of cultures, ethnic groups, religions and built form. That heritage remains no less evident today in a diversity which ranges from the highly developed and managed urban environment of Singapore, to the historic relicts of Malay, Dutch and British power in Malacca, to the contrast of native *praus* bobbing alongside the giant tankers that criss-cross the region's ports. If only a part of the fascination of seeking to unravel that diversity has been conveyed in this book it will have more than served its purpose.

REFERENCES

Abdullah, A.R., Woon, W.C. & Bakar, R.A. (1996). Distribution of oil and grease and petroleum hydrocarbons in the Straits of Johor, Peninsular Malaysia. *Bulletin of Environmental Contamination and Toxicology*, 57: 155–162.

Abdullah, C., Syahrul, J. & Mukhtar, S. (1995). The analysis of heavy metals in the sediment water bodies around Sungei pakning, Bengkalis District, Riau Province, Indonesia. In: D. Watson, K.S. Ong & G. Vigers (Eds), *Advances in marine environmental management and human health protection*. Proceedings of the ASEAN-Canada Midterm Technical Review Conference on Marine Science, 24–28 October 1994, Singapore, 276–278.

Abdullah, S. (1992). Coastal erosion in Malaysia: problems and challenges. In: H.D. Tjia & M.S. Abdullah (Eds), *The coastal zone of Peninsular Malaysia*. Universiti Kebangsaan Malaysia, Bangi, 79–92.

Affifuddin, O. (1975). *Implementation of rural development: institution building in the Muda Region*. MADA Monograph 27, MADA, Alor Star, Malaysia, 32.

Ahmad, J. & Ismail, M.N. (1972). Occurrence of beach rocks and coral heads in the Langkawi island group. *Sains Malaysiana*, 1: 113–114.

Ahmat, S. (1965). American trade with Singapore, 1819–65. *Journal of the Malay Branch, Royal Asiatic Society*, 38, 2: 241–257.

Akira, M. (1990). Review of actual situation on sewage works in Malaysia. Paper presented at the National Seminar on Wastewater Treatment Systems, 12–15 November, Johor Bahru, Ministry of Housing and Local Government.

Aleva, G.J.J., Bon, E.H., Nossin, J.J. & Sluiter, W.J. (1973). A contribution to the geology of the part of the Indonesian Tinbelt: the sea areas between Singkep and Bangka islands and around the Karimata islands. *Bulletin of the Geological Society of Malaysia*, 6: 257–271.

Allen, G. and Donnithorne, A. (1957). *Western enterprise in Indonesia and Malaya*. London: Allen and Unwin.

Andaya, B.W. & Andaya, L. (1988). *A history of Malaysia*. Basingstoke: Macmillan.

Anderson, J.A.R. (1964). The structure and development of the peat swamps of Sarawak and Brunei. *Journal of Tropical Geography*, 18: 7–16.

Andriesse, J.P. (1974). *Tropical lowland peat in Southeast Asia*. Department of Agriculture Research of the Royal Tropical Institute, Amsterdam. Communication 63.

Ang, K.J. (1990). Status of aquaculture in Malaysia. In: M.M. Joseph (Ed.), *Aquaculture in Asia*. Asian Fisheries Society, Indian Branch, 265–279.

Ariyathavaratnam, K. (1989). Development of a sewerage system for the Federal Territory of Kuala Lumpur. Paper presented at seminar on Domestic Wastewater Treatment Alternatives, 9–11 November 1987, University Pertanian, Malaysia.

Aw, P.C. (1986). Geology and exploitation of kaolin deposits in the Bidor area, Peninsular Malaysia. *GEOSEA V, Proceedings of the Geological Society of Malaysia Volume II*, 20: 601–617.

Barbier, E.B. (1993). Sustainable use of wetlands—valuing tropical wetland benefits: methodologies and applications. *The Geographical Journal*, 159, 1: 22–32.

Bastin, J. & Roolvink, R. (Eds) (1964). *Malayan and Indonesian studies*. Oxford: Clarendon Press.

Batchelor, B.C. (1979a). Geological characteristics of certain coastal and offshore placers as essential guides for tin exploration in Sundaland, Southeast Asia. *Bulletin of the Geological Society of Malaysia*, 11: 283–313.

Batchelor, D.A.F. (1979b). Discontinuously rising Late Cainozoic eustatic sea levels, with special reference to Sundaland, Southeast Asia. *Geologie Minjb*, 58: 1–20.

Batchelor, D.A.F. (1988). Dating of Malaysian fluvial tin placers. *Journal of Southeast Asian Earth Sciences*, 2: 3–14.

Beckman, R.C., Grundy-Warr, C. & Forbes, V.L. (1994). *Acts of piracy in the Malacca and Singapore Straits*. Durham: International Boundaries Research Unit, Department of Geography, University of Durham.

Bemmelen, R.W. van (1967). *The geology of Indonesia: Volume 1A, General geology of Indonesia and adjacent archipelagos*. The Hague: Government Printing Office.

Bemmelen, R.W. van (1949). *The geology of Indonesia*. The Hague: SDU.

Bird, E.C.F. & Barson, M.M. (1977). Measurement of physiographic changes on mangrove-fringed estuaries. *Journal of Tropical Geography*, 11, 8–17

Bird, E.C.F & Teh T.S. (1990). Current state of the coastal zone in Malaysia. *Malaysian Journal of Tropical Geography*, 21, 1: 9–24.

Biswas, B. (1973). Quaternary changes in sea level in the South China Sea. *Bulletin of the Geological Society of Malaysia*, 6: 229–256.

Booth, A. (1988). *Agricultural development in Indonesia*. Sydney: Allen and Unwin.

Bosch, J.H.A. (1988). Quaternary geological map of Peninsular Malaysia. Geological Survey of Peninsular Malaysia. Geological Survey Malaysia (1st Edn), Scale 1:1,000,000.

Bosch, J.H.A. (1989). The Quaternary deposits in the coastal plains of Peninsular Malaysia. *Geological Survey Malaysia*, Report QG/1.

Boxer, C.R. (1965). *The Dutch seaborne empire, 1600–1800*. London: Hutchinson.

Braddell, R. (1937). An introduction to the study of ancient times in the Malay peninsula and the Straits of Malacca: Pre-Funan. *JMBRAS*, 15, 3: 108–170.

Braddell, R. (1939). An introduction to the study of ancient times in the Malay Peninsula and the Straits of Malacca: Pre-Funan; Adenda. *JMBRAS*, 17, 1: 171–237.

Bradford, E.F. (1972). Geology and mineral resources of the Gunung Jerai area, Kedah. *Geological Survey Malaysia*, Memoirs 13.

Braudel, F. (1975). *The Mediterranean and the Mediterranean world in the age of Phillip II*. 2 vols., London: Collins.

Breman, J. (1989). *Taming the coolie beast. Plantation society and the colonial order in Southeast Asia*. Delhi: Oxford University Press.

Brookfield, H., Potter, L. & Byran, Y. (1995), *In place of the forest*. Tokyo: Tokyo's United Nations University Press.

Brown, F.B. (1972). Land and water resources development in South East Sumatra. Agronomy. *Food and Agriculture Organisation—Government of Indonesia*, Department of Agriculture.

Bruijn, J.R. (1980). Between Batavia and the Cape: shipping patterns of the Dutch East India Company. *Journal of Southeast Asian Studies*, 11, 2: 251–265.

Brzeski, V. & Newkirk, G. (1997). Integrated coastal food production systems: a review of current literature. *Ocean and Coastal Management*, 34, 1: 55–71.

Bucholz, H.J. (1987). *Law of the sea zones in the Pacific Ocean*. Singapore: Institute of South East Asian Studies.

Burbridge, P.R. (1988). Coastal and marine resource management of the Strait of Malacca. *Ambio*, 17, 3: 170–177.

Burnham, C.P. (1978). Soil formation and its variation with altitude in Malaya and western Sabah. PhD thesis, Wye College, University of London.

Burnham, C.P. (1989). Pedological processes and nutrient supply from parent material in tropical soils. In: J. Proctor (Ed.), *Mineral nutrients in tropical forest and savanna ecosystems* (pp. 27–41). Special Publications No. 9, The British Ecological Society. Oxford: Blackwell Scientific Publications.

Business Times (1999) Do more for Straits safety, Japan urges user nations. 28 January.

Cable, B. (1937). *A hundred year history of the P and O*. London: Nicholson and Watson.

Castle, L. & Findlay, C. (Eds) (1988). *Pacific trade in services*. Sydney: Allen and Unwin.

Challis, N. (1996). Financing the development and operation of shore reception facilities: case studies. Regional Conference on Sustainable Financing Mechanisms for the Prevention and Management of Marine Pollution: Public Sector–Private Sector Partnerships, 14–16 November 1996, Manila. Also published in *Tropical Coasts*, 1996, 3, 2: 19–21.

Chaudhuri, K. (1978). *The trading world of Asia and the East India Company*. Cambridge: Cambridge University Press.

Cheah, K.W. (1997). The present status and future directions of the maritime industry in Malaysia. *MIMA Bulletin*, 6, 2: 1–9.

Cheah, U.B & Lum, K.Y. (1993). Pesticide contamination in the environment. In: B.G. Yeoh *et al* (Eds), *Waste management in Malaysia: current status and prospects for bioremediation* (pp. 169–176), Malaysia: Ministry of Science, Technology and the Environment.

Cheong, L. (1990). Aquaculture development in Singapore. In: M.M. Joseph (Ed.), *Aquaculture in Asia* (pp. 325–332). Asian Fisheries Society, Indian Branch.

Chew T.Y. (1993). Tapioca processing wastewater. In: B.G. Yeoh *et al* (Eds), *Waste management in Malaysia: current status and prospects for bioremediation* (pp. 153–158). Malaysia: Ministry of Science, Technology and the Environment.

Chia, L.S. (1992). *Singapore's urban coastal area: strategies for management*. Manila: ASEAN/US Coastal Resources Management Project, Technical Publications Series 9.

Chia, L.S. (1995). Protecting the marine environment of ASEAN from ship-generated oil pollution and Japan's contribution to the region. *Visiting Research Fellow Monograph Series*, 245: 141. Tokyo: Institute of Developing Economies.

Chia, L.S. (1997). Alternative routes for oil tankers: a financial, technical and economic analysis. In: A. Hamzah (Ed.), *The Straits of Malacca: international co-operation in trade, funding and navigational safety* (pp. 103–122). Kuala Lumpur: Pelanduk Publications, Maritime Institute of Malaysia.

Chia, L.S. & MacAndrews, C. (Eds) (1981). *Southeast Asian seas: frontiers for development*. Singapore: Institute of Southeast Asian Studies.

Chia, L.S., Habibullah K. & Chou, L.M. (1988). *The coastal environmental Profile of Singapore*. Manila: ASEAN/US Coastal Resources Management Project, Technical Publications Series 3.

Cho, W.K. (1995). Coastal zone in Malaysia—issues and management (6). Paper presented at the ESCAP/ADB/UNEP Workshop on Coastal and Marine Environmental Management, 27–29 March 1995, Bangkok, Thailand.

Chong, P.L., Hambal, H., Zuridah, O.M. & Nagaraj, G. (1987). Aquaculture development in Malaysia. Country paper, Seminar on Aquaculture Development in Southeast Asia, SEAFDEC, Iloilo, Philippines, 8–13 September 1987.

Chou, L.M., Wilkinson, C.R., Licuanan, W.R.Y., Alino, P., Cheshire, A.C., Loo, M.G.K., Tangjaitrong, S., Sudara, S., Ridzwan, A.R. & Soekarno (1994). Status of coral reefs in the ASEAN region. In: C.R. Wilkinson, S. Sudara & L.M. Chou (Eds), *Proceedings of the Third ASEAN-Australia Symposium on Living Coastal Resources. Volume I: Status reviews* (pp. 1–10). Chulalongkorn University, Thailand, 16–20 May, Australian Institute of Marine Science, Townsville.

Chua, T.-E., Adrian Ross, S. & Huming, Y. (Eds) (1997). *Malacca Straits: environmental profile.* Quezon City, Philippines: GEF/UNDP/IMO Regional Programme for the Prevention and Management of Marine Pollution in the East Asian Seas.

Chua, T.-E. (1995). Marine pollution: developments since UNCLOS III and prospects for regional cooperation. In: K.L. Koh, R.C. Beckman & Chia, Lin Sien (Eds), *Sustainable development of coastal and ocean areas in Southeast Asia: post-Rio perspectives* (pp. 144–176). Singapore: National University of Singapore.

Cicin-Sain, B. & Knecht, R.W. (1998). *Integrated coastal and ocean management: concepts and practices.* Washington: Island Press.

Cleary, M. (1996). Indigenous trade and European economic intervention in North-West Borneo, c. 1860–1930. *Modern Asian Studies,* 30, 2: 301–324.

Cleary, M. (1997). From hornbills to oil? Patterns of indigenous and European trade in colonial Borneo. *Journal of Historical Geography,* 23, 1: 29–45.

Cleary, M. & Eaton, P. (1996). *Tradition and reform: land tenure and rural development in Southeast Asia.* Kuala Lumpur: Oxford University Press.

Cleary, M. & Shaw, B. (1994). Ethnicity, development and the New Economic Policy: the experience of Malaysia, 1971–1990. *Pacific Viewpoint,* 35, 1: 83–106.

Cobbing, E.J., Mallick, D.I.S., Pitfield, P.E.J. & Teoh, L.H. (1986). The granites of the Southeast Asian tin belt. *Geological Society of London,* 143: 537–550.

Coedes, G. (1966). *The making of Southeast Asia.* Berkeley, California: University of California Press.

Collins, N.M., Sayer, J.A. & Whitmore, T.C. (Eds) (1991). *The conservation atlas of tropical forests: Asia and the Pacific.* London: Simon & Schuster.

Courtenay, P. (1980). *Plantation agriculture.* London: Bell and Hyman.

Cowan, M. (1950). Early Penang and the rise of Singapore. *JMBRAS,* XXIII, II: 3–207.

Curray, J.R. (1961). Late Quaternary sea level: a discussion. *Bulletin of the American Geological Society,* 72: 1707–1712.

Dahuri, R. & Pahlevi, R.Z. (1994). North Sumatra, Indonesia. In: S. Holmgren (Ed.), An environmental assessment of the Bay of Bengal region. *Bay of Bengal Report,* 13–32.

Daniel, J.G. & Kulasingam, A. (1974). Problems arising from large scale jungle clearing for agricultural use: Malaysian experience. *Malaysian Forester,* 37: 152–160.

Darby, H.C. (1932). The Medieval sea state. *Scottish Geographical Magazine,* 48: 136–149.

Dayan D. & Sjafrizal, (1989). Acheh: the LNG boom and enclave development. In: H. Hill (Ed.), *Unity and diversity: regional economic development in Indonesia since 1970* (pp. 107–124). Singapore: Oxford University Press.

Demets, C., Gordon, R.G., Argus, D.F. & Stein, S. (1990). Current plate motions. *International Journal of Geophysics,* 101: 425–478.

Department of Environment (1989). *Environmental quality report 1988.* Malaysia: Ministry of Science, Technology and the Environment.

Department of Environment (1996). *Malaysia environmental quality report 1995.* Malaysia: Ministry of Science, Technology and the Environment.

Department of Fisheries (1995). *Annual fisheries statistics.* Ministry of Agriculture, Kuala Lumpur, Malaysia.

Dicken, P. (1992). *Global shift.* London: Paul Chapman.

Dobby, E. H. (1950). *Southeast Asia.* London: University of London Press.

Donner, W. (1987). *Land use and environment in Indonesia.* London: C. Hurst & Co.

Douglas, I. (1967). Erosion of granite terrains under tropical rain forest in Australia, Malaysia and Singapore. Extract from the Symposium on River Morphology, General Assembly of Bern, September–October 1967, 31–39.

Dow, K. (1997). An overview of pollution issues in the Straits of Malacca. In: A. Hamzah (Ed.), *The Straits of Malacca: international co-operation in trade, funding and navigational safety* (pp. 61–102). Kuala Lumpur: Pelanduk Publications, Maritime Institute of Malaysia.

Driessen, P.M. (1978). Peat soils. In: *Soils and rice* (pp. 763–779). Los Banos: International Rice Research Institute.

Dunn, F.L. (1975). *Rainforest collectors and traders. a study of resource utilisation in modern and ancient Malaya.* Malaysian Branch of the Royal Asiatic Society, Monograph 5.

Dwyer, D. (Ed.) (1988). *South East Asian development: geographical perspectives.* London: Longman.

Emmel, F.J. & Curray, J.R (1982). A submerged Late Pleistocene delta and other features related to sea level changes in the Malacca Strait. *Marine Geology,* 47: 197–216.

Esterle, J.S. & Ferm, J.C. (1994). Spatial variability in modern tropical peat deposits from Sarawak, Malaysia and Sumatra, Indonesia: analogues for coal. *Coal Geology,* 26: 1–41.

Eswaran, H. & Wong, C.B. (1978). A study of deep weathering of granite in Peninsular Malaysia. *Journal of the Soil Science Society of America,* 42: 144–158.

Finn, D.P. *et al* (1979). *Oil pollution from tankers in the Straits of Malacca: a policy and legal analysis.* Honolulu: East-West Centre, Open Grants Office.

Flenley, J.R. (1988). Palynological evidence for land use changes in South-East Asia. *Journal of Biogeography,* 15: 185–197.

Fournier, F. (1960). *Climat et Erosion: la relation entre l'erosion du sol par l'eau et precipitations atmospheriques.* Paris: PUF.

Frankel, E.G. (1995). *Ocean environmental management: a primer on the role of the oceans and how to maintain their contributions to life on earth.* New Jersey: Prentice Hall.

Freestone, D. & Penthick, J. (1994). Sea level rise and maritime boundaries: international implications of impacts and responses. In: G.H. Blake (Ed.), *Maritime boundaries, world boundaries, Volume 5.* London: Routledge.

Furnivall, J.S. (1939). *Netherlands India: a study of plural economy.* Cambridge: Cambridge University Press.

Gasparon, M. & Varne, R. (1995). Sumatran granitoids and their relationship to Southeast Asian terranes. *Tectonophysics,* 251: 277–299.

Geyh, M.A., Kudrass, H.R. & Streif, H. (1979). Sea level changes during the Late Pleistocene and Holocene in the Strait of Malacca. *Nature,* 278: 441–443.

Gibson-Hill, C. (1950). The Indonesian Trading Boats reaching Singapore. *Journal of the Malayan Branch, Royal Asiatic Society,* 23, 1: 108–138.

Goh, B. (1997). Impact of pollution on marine biodiversity. *Tropical Coasts,* 4, 1: 20–22.

Gorbett, D.J. & Hutchison, C.S. (Eds) (1973). *Geology of the Malay Peninsula.* Chichester: Wiley-Interscience.

Gold, E. (undated). *Transit services in international straits: towards shared responsibilities?* Kuala Lumpur: Malaysian Institute of Maritime Affairs Issue Paper.

Government of Malaysia and Asian Development Bank (1987). *Klang Valley Environmental Improvement Project.* Prepared by Engineering Science and Satec International and Department of the Environment.

Grace, L.M., Woo, K.H. & Chou, L.M. (1987). Singapore country/status report. In: Development and Management of Living Marine Resources Workshop on Pollution and Other Ecological Factors in Relation to Living Marine Resources (pp. 188–274). Report of ASEAN-Canada Cooperative Programme on Marine Science.

Gupta A., Rahman, A. & Wong, P.P. (1987). The old alluvium of Singapore and the extinct drainage system to the South China Sea. *Earth Surface Processes and Landforms,* 12: 259–275.

Gupta, B.S., Poulose, T.T. & Bhatia, H. (1974). *The Malacca Straits and the Indian Ocean: a study of the strategic and legal aspects of a controversial sea-lane.* Delhi: Macmillan.

Haile, N.S. (1971) Quaternary shorelines in West Malaysia and adjacent areas of the Sunda shelf. *Quaternaria,* 15: 333–343.

Haile, N.S. (1975) Postulated late Cainozoic high sea levels in the Malay Peninsula. *JMBRAS,* 68: 1: 78–88.

Hall, D.E. (1987). *A History of SE Asia* (4th edn). London: Macmillan.

Hamdan, J. & Burnham, C.P. (1996). The contribution of nutrients from parent material in three deeply weathered soils of Peninsular Malaysia. *Geoderma,* 74: 219–233.

Hamilton, W. (1979). Tectonics of the Indonesia region. *US Geological Survey Profiles,* 1078: 345.

Hamzah, A. (1996). Providing for safer and cleaner seas in the Straits of Malacca. Regional Conference on Sustainable Financing Mechanisms for Marine Pollution Prevention and Management: Public Sector–Private Sector Partnerships, 14–16 November 1996. Also published in *Tropical Coasts,* 3, 2: 7–11.

Hamzah, A. (1997). Global funding for navigational safety and environment protection. In: A. Hamzah (Ed.), *The Straits of Malacca: international co-operation in trade, funding and navigational safety* (pp. 125–144). Kuala Lumpur: Pelanduk Publications, Maritime Institute of Malaysia.

Hanson, A.J. & Koesoebiono (1979). Settling coastal swamplands in Sumatra: A case study for integrated resource management. In: C. MacAndrews & L.S. Chia (Eds), *Developing economies and the environment: the Southeast Asian experience* (pp. 121–175). Singapore: McGraw-Hill.

Hantoro, W.S., Faure, H., Djuwansah, R., Faure-Denard, L. & Pirazzoli, P.A. (1995). The Sunda and Sahul continental platform: lost land of the last glacial continent in S.E. Asia. *Quaternary International,* 29, 30: 129–134.

Hanus, V., Spichak, A. & Vanek, J. (1996). Sumatran segment of the Indonesian subduction zone: morphology of the Wadati-Benioff zone and seismotectonic pattern of the continental wedge. *Journal of Southeast Asia Earth Science,* 13, 1: 39–60.

Hardjono, J.M. (1977). *Transmigration in Indonesia.* Singapore: Oxford University Press.

Harrison, B. (1953). Trade in the Straits of Malacca in 1785, *Journal of the Malay Branch, Royal Asiatic Society,* 26, 1: 56–62.

Hattendorf, J. (Ed.) (1996). *Maritime history: 1: the age of discovery.* Malabar, Florida: Krieger.

Hew, S.T. (1984). Gravel pump tin mining in Malaysia [in Chinese]. Kuala Lumpur: Nanyang Muda Sdn Bhd.

Hill, H. (Ed.) (1989). *Unity and diversity.* Kuala Lumpur: Oxford University Press.

Ho, K.C. (1993). Industrial restructuring and the dynamics of city-state adjustments. *Environment and Planning*, A, 25: 47–62.

Ho, N.K. (1994). Integrated weed management of rice in Malaysia: some aspects of the Muda Irrigation Scheme's approach and experience. In: S.S. Sastroutomo & A.A. Bruce (Eds), *Appropriate weed control in Southeast Asia* (pp. 83–97). Kuala Lumpur: CAB International.

Ho, N.K., Asna, B.O., Aznan, A. & Rabirah, A. (1990). Herbicide usage and associated incidences of poisoning in the Muda area, Malaysia—a case study. Proceedings of the 3rd Tropical Weed Science Conference, Kuala Lumpur, 321–333.

Ho Y.-C. (1987). Control and management of pollution of inland waters in Malaysia. *Arch. Hydrobiol. Beih. Ergebn. Limnol. Stuttgart*, 28: 547–556.

Holdgate, D. (1992). Opening address. In: K. Sherman and T. Laughlin (Eds), *Large marine ecosystems concept and its application to regional marine resource management* (pp. 19–24). Switzerland: Marine Conservation and Development Reports, IUCN.

Huat, K.K. (1978). Implementation of regulation for domestic fishermen. In: F.T. Christy, (Ed.), *Law of the sea: problems of conflict and management of fisheries in Southeast Asia*. Manila: ICLARM/ISEAS Conference Proceedings, No. 2, 42–48.

Hutchison, C.S. (1983). *Economic deposits and their tectonic settings*. London: MacMillan.

Hutchison, C.S. (Ed.) (1996) *Geological evolution of South-East Asia*. Kuala Lumpur: Geological Society of Malaysia.

Huxley, T. (1996). Southeast Asia in the study of international relations. *Pacific Review*, 9, 2: 199–228.

Hyde, F. (1973). *Far Eastern trade, 1860–1914*. London: A & C Black.

Idriss, A. (1990). *Malaysia's New Economic Policy*. Petaling Jaya: Pelanduk Publications.

Idris, S.M. (1977). Opening address. In: *Malaysian fisheries—a diminishing resource* (pp. 2–3).

Institute of Southeast Asian Studies (1999). *Regional outlook 1999–2000*. Singapore: ISEAS.

International Centre for Living Aquatic Resource Management (1991). *The coastal environmental profile of South Johore, Malaysia*. Manila: ASEAN/US Coastal Resources Management Project, Technical Publications Series 6.

Ismail I. (1993). Coastal resources of the Malacca Strait and their development. Paper presented at the National Conference on the Straits of Malacca, 11 November 1993, Kuala Lumpur.

Jaafar, A.B. & Harun, H. (1979). Analysis of water pollution complaints in Peninsular Malaysia, 1978. Kuala Lumpur: Division of Environment. Unpublished.

Jesusadon, J.V. (1990). *Ethnicity and the economy: the state, Chinese business and multinationals in Malaysia*. Singapore: Oxford University Press.

JICA (1981). *Master plan for sewerage and drainage system project Kelang, Port Kelang and its environs*. Majlis Perbandaran Kelang. Japan International Cooperation Agency.

Jobson, D.H., Boulter, C.A. & Foster, R.B. (1994). Structural controls and genesis of epithermal gold-bearing breccias at the Lebong Tandai mine, Western Sumatra, Indonesia. *Journal of Geochemical Exploration*, 50: 409–428.

Jones, C.R. (1981) Geology and mineral resources of Perlis, North Kedah and the Langkawi Islands. *Geological Survey Malaysia*, Memorandum 17.

Jordan, C.F. (1985). *Nutrient cycling in tropical forest ecosystems*. Chichester: Wiley.

Kamaludin, H. (1993). The changing mangrove shoreline in Kuala Kurau, Peninsular Malaysia. In: C.D. Woodroffe (Ed.), Late Quaternary evolution of coastal and lowland riverine plains of Southeast Asia and Northern Australia. *Sedimentary Geology*, 83: 187–197.

Katili, J.A. (1975). Vulcanism and plate tectonics in the Indonesian island arcs. *Tectonophysics*, 26: 165–168.

Katili, J.A. (1985). *Advancement of geosciences in the Indonesian region* (pp. 1–248). Bandung, Indonesia: The Indonesian Association of Geologists.

Katili, J.A. & Hehuwat, F. (1967). On the occurrence of large transcurrent faults in Sumatra, Indonesia. *Journal of Geosciences*, 10: 5–17.

Keller, G.H. & Richards, A.F. (1967). Sediments of the Malacca Strait, Southeast Asia. *Journal of Sedimentary Petrology*, 37: 102–127.

Kemp, P. (1980). *Encyclopaedia of ships and shipping*. London: Stanford Maritime.

Keosoemadinata, R.P & Nelson, V.E. (1970). Mineral resources in Indonesian development. In: H.W. Beers (Ed.), *Indonesia: resources and their technical development* (pp. 117–139). Lexington: University of Kentucky Press.

Khairulmaini, O.S. (1994). Geomorphological and bathymetrical considerations in the identification and ranking of potential power station sites in Peninsular Malaysia. *Geoforum*, 25, 3: 381–399.

Khalil H. (1985). Aquaculture programme for fishermen (pp. 18–21). Report on the Aquaculture Conference, 9–12 December, Johor Bahru, Malaysia.

Khoo, T.T. (1996). Geomorphological evolution of the Merbok estuary and its impact on the early state of Kedah, northwest peninsular Malaysia. *Journal of SE Asian Earth Sciences*, 13, 3–5: 347–371.

Khoo, T.T. & Tan, B.K. (1983). Geological evolution of Peninsular Malaysia. *Proceedings of the Workshop on Stratigraphic Correlation of Thailand and Malaysia*, 1: 253–290.

Kirby, S.W. (1928). Johore in 1926. *Geographical Journal*, 71: 240–260.

Klinken, G. van & Halim, Q.A. (1982). Offshore Gunung Jerai shallow seismic survey. *Bulletin of the Geological Society of Malaysia*, 15: 71–82.

Koe, L.C.C. & Aziz, M.A. (1995). Suggested approach to control marine pollution from land-based sources. Paper presented at the Asia-Pacific Regional Conference on the Prevention of Marine Pollution from Land and Sea Sources, 4–7 December 1995, Germany–Singapore Environmental Technology Agency, Singapore.

Koopmans, B.N. (1964). Geomorphological and historical data of the lower course of the Perak River (Dindings). *JMBRAS*, 37, 2: 175–191.

Kudrass, H.R. & Schluter, H.U. (1994). Development of cassiterite-bearing sediments and their relation to Late Pleistocene sea-level changes in the Straits of Malacca. *Marine Geology*, 120: 175–202.

Kumar, S. & Siddique, S. (1994). Beyond economic reality: new thoughts on the growth triangle. *Southeast Asian Affairs*, 47–56.

Lamb, A. (1980). Pengkalen Bujang: an ancient port in Kedah. *Lembah Bujang*, 79–81, Persatuan Sejarah Melayu (Malay Historical Society).

Larsson, J., Folke, C. & Kautsky, N. (1994). Ecological limitations and appropriation of ecosystem support by shrimp farming in Columbia. *Environmental Management*, 18: 663–676.

Law, A.T. (1980). Sewage pollution in Kelang River and its estuary. *Pertanika* 3, 1: 13–19.

Law, A.T. (1984). Monitoring of sewage pollution in the estuarine and coastal water of Port Kelang, Malaysia. Proceedings of the International Conference on Development and Management of Living Aquatic Resources, University Pertanian Malaysia, Kuala Lumpur.

Law, A.T. & Azhar, O. (1985). Fecal *coliform* bacteria distribution in the coastal waters of Port Dickson. *Pertanika*, 8, 1: 131–134

Law, A.T. & Singh, A. (1987). Distribution of mercury in the Kelang Estuary. *Pertanika*, 10: 175–181.

Law, A.T. & Singh, A. (1991). Relationships between heavy metals content and body weight of fish from the Kelang Estuary, Malaysia. *Marine Pollution Bulletin*, 20, 2: 86–89.

Lee, B.S. & Ong, S.H. (1983). Problems associated with pesticide use in Malaysia. Proceedings of the International Symposium on Pesticide Use in Developing Countries—present and future, 25–35.

Lee, S.K., Tan W. & Havanond, S. (1996) Regeneration and colonisation of mangrove on clay-filled reclaimed land in Singapore. *Hydrobiology*, 319: 23–35.

Lee, T.Y & Lawyer, L.A. (1995). Cenozoic plate reconstruction of Southeast Asia. *Tectonophysics*, 251: 85–138.

Lee, W.B. (1994). Maritime disasters and casualties in the Straits of Malacca. Paper presented at the 1st International Conference on Maritime Crisis Management: The Aftermath of Maritime Disasters, 26–28 September 1994, Kuala Lumpur.

Lee, Y.S.F. (1997). The privatisation of solid waste infrastructure and services in Asia. *Third World Planning Review*, 19, 2: 139–161.

Leekpai, C. (1991). The value of coastal resources in national economic development. In: T.E. Chua & L.E. Scura (Eds.), *Managing ASEAN's coastal resources for sustainable development: role of policy makers, scientists, donors, media and communities*. Manila: ICLARM Proceedings.

Leifer, M. (1978). *International straits of the world—Malacca, Singapore and Indonesia*. Netherlands: Alphen aan den Rijn: Sijthoff & Noordhoff Publishers.

Leigh, C.H. (1978). Slope hydrology and denudation in the Pasoh forest reserve. I—Surface wash: experimental techniques and some preliminary results. *Malayan Nature Journal*, 30: 179–197.

Leinbach, T.R. (1972) The spread of modernization in Malaya: 1895–1969. *Tijdschrift voor Economische en Sociale Geografie*, 63: 262–277.

Lewis, D. (1970). The growth of the country trade to the Straits of Malacca, 1760–1777. *Malay Branch, Royal Asiatic Society*, 43, II: 114–130.

Lewis, D. (1995). *Jan Compagnie in the Straits of Malacca, 1641–1795*. Athens: Ohio University Centre for International Studies.

Lim, B.C. (1993). Geotechnical studies for the rehabilitation of mining land. *Pegama*, 3, 1: 3–6.

Lim, C.H. (1992). Reclamation of peat land for agricultural development in West Johor. In: B.Y. Aminuddin (Ed.), *Tropical peat*. Proceedings of the International Symposium on Tropical Peatland, 6–10 May 1991, Kuching, Malaysian Agricultural Research and Development Institute, Kuala Lumpur, 177–189.

Lim, K.H. (1987). Trials on long term effects of application of POME on soil properties, oil palm nutrition and yields. Proceedings of the 1987 International Oil Palm/Palm Oil Conference—Agriculture, Kuala Lumpur, 575–595.

Lim, K.H., Quah, S.K., Gillies, D. & Wood, B.J. (1984). Palm oil effluent treatment and utilization in Sime Darby Plantations—the current position. Proceedings of the Workshop Palm Oil Research Institute Malaysia, 9: 42–52.

Lim, M.H. (1985). Contradictions in the development of Malay capital: state accumulation and legitimation. *Journal of Contemporary Asia*, 15, 1: 37–64.

Lloyds of London (1997). *Ports of the world*. London: Lloyds.

Lobbrecht, A.H., Mak, W., van de Kerk, F. & Beeker, A. (1985). *Swamp land development in Indonesia: an inventorial study of the alluvial plains of Sumatera, Kalimantan and Sulawesi and their development related to transmigration policies*. Delft: Department of Engineering, Delft University of Technology.

Lokman, M.H. & Othman, F.H. (1991). A case study of Klang valley and Johor Bahru, Malaysia. Paper presented at the Enviroworld 1991 Conference on Solid and Hazardous Waste Management, June, Singapore.

Low, J.K.Y. & Chou, L.M. (1994). Fish diversity of Singapore mangroves and the effect of habitat management. In: S. Sudara, C.R. Wilkinson & L.M. Chou (Eds), *Proceedings of the Third ASEAN–Australia Symposium on Living Coastal Resources. Volume 2: Research papers* (pp. 465–469). Bangkok: Department of Marine Sciences, Chulalongkorn University.

Ma, A.N. & Ong, A.S.H. (1987). Potential biomass energy from palm oil industry. *PORIM Bulletin*, 14: 10–15.

Ma, A.N., Cheah, S.H. & Chow, M.C. (1993). Current status of oil palm processing wastes management. Waste management in Malaysia. In: B.G. Yeoh *et al* (Eds), *Current status and prospects for bioremediation* (pp. 111–136). Kuala Lumpur: Ministry of Science, Technology and the Environment.

MACA (1989). *Annual report 1988/89*. Malaysian Agricultural Chemicals Association.

Macdonald, A. & Anderson, N. (1996). Marine electronic highway. Regional Conference on Sustainable Financing Mechanisms for Marine Pollution Prevention and Management: Public Sector–Private Sector Partnerships, 14–16 November 1996. Also published in *Tropical Coasts*, 3, 2: 15–18.

Maheswaran, A. & Singam, G. (1977). Pollution control in the palm oil industry—promulgation of regulations. *Planter*, 53: 470–476.

Majid, S.A. (1988). *Fishing industry in Asia and the Pacific*. Tokyo: Asian Productivity Organization.

Majlis Perbandaran Seberang Perai (Seberang Perai City Council) (1981). Final engineering report. Penang: Butterworth/Bukit. Mertajam Metropolitan Sewerage Project.

Manguin, P. (1980). The Southeast Asian ship: an historical approach. *Journal of Southeast Asian History*, 11, ii: 266–276.

Marr, J.C. (1981). Southeast Asian marine fishery resources and fisheries. In: C. MacAndrews & L.S. Chia (Eds), *Southeast Asian seas: frontiers for development* (pp. 75–109). Singapore: McGraw-Hill.

Mastaller, M. (1996). *Mangroves—the forgotten forest between land and sea*. Kuala Lumpur: Tropical Press.

McCabe, R., Celaya, M., Cole, J.T., Han, H.C., Ohnstad, T., Paijitprapapon, P. & Thitipawarn, V. (1988). Extension tectonics: the Neogene opening of the north–south trending basins of central Thailand. *Journal of Geophysical Research*, 93, 11: 899–910.

McIntyre, W.D. (1967). *The imperial frontier in the tropics, 1865–75*. London: Macmillan.

McIntyre, W.D. (1979). *The rise and fall of the Singapore naval base, 1919–1942*. Basingstoke: Macmillan.

McKinnon, E.E. (1984). New data for studying the early coastline in the Jambi area. *SPAFA Digest*, 5, 2: 4–8.

Meilink-Roelofsz, M. (1962). *Asian trade and European influence in the Indonesian archipelago between 1500 and about 1630*. The Hague: Martinus Nijhoff.

Menasveta, D. (1994). Fisheries management in the exclusive economic zones of Southeast Asia before and after Rio and prospects for regional cooperation. In: K.L. Koh, R.C. Beckman & L.S. Chia (Eds), *Sustainable development of coastal and ocean areas in Southeast Asia* (pp. 98–134). Singapore: Law faculty, National University of Singapore.

Meng, Q. (1987). *Land-based marine pollution: international law development*. London: Graham & Trotman/Martinus Nijhoff.

Merican, A.B.O. (1977). The status of the Malaysian fisheries management and development aspects. In: *Malaysian fisheries—a diminishing resource* (pp. 21–28). Penang: Consumers Association of Penang.

Metcalfe, I.. (1988). Southeast Asia. In: C. Diaz (Ed.), *The carboniferous of the world.* Madrid: I.U.G.S. Publication No. 16, Instituto Geologico y Minero de Espana.

Meyer, F.V. (1948). *Britain's colonies in world trade.* London: Oxford University Press.

Meyerhoff, A.A. (1995). Surge-tectonic evolution of Southeastern Asia: a geohydrodynamics approach. *Journal of Southeast Asian Earth Sciences,* Memorial Issue, 12, 3/4: 145–247.

Ministry of Environment, Singapore (1989). Country report on the assessment of pollution from land-based sources and their impact on the marine environment. Country report presented at the UNEP-COBSEA seminar on the Assessment of Pollution from Land-based Sources and Their Impact on the Environment, 25–27 January 1989, Singapore.

Mohd, T., Hamdan, A.B., Ahmad Tarmizi, M. & Roslan, A. (1987). Recent progress on research and development on peat for oil palm. *PORIM Bulletin,* 34: 11–35.

Mohd Zahari, A.B., Abd Wahab, N., Ting, C.C. & Abd Rahim, M. (1981). Distribution, characterisation and utilisation of problem soils in Malaysia—a country report. Proceedings of the Symposium on Distribution, Characterisation and Utilisation of Problem Soils, Tsukuba, Ibaraki, Japan.

Morner, N.A. (1971), The position of ocean levels during the interstadial at about 20,000 BP: a discussion from a climatic glaciologic point of view. *Canadian Journal of Earth Science,* 8: 132–143.

Muhammad Razif Ahmad (1997). The financial cost of risk management in the Straits of Malacca. In: A. Hamzah (Ed.), *The Straits of Malacca: international co-operation in trade, funding and navigational safety* (pp. 187–219). Kuala Lumpur: Pelanduk Publications, Maritime Institute of Malaysia.

Mutalib, A.A., Lim, J.S., Wong, M.H. & Koonvai, L. (1992). Characterization, distribution and utilisation of peat in Malaysia. In: B.Y. Aminuddin (Ed.), *Tropical peat.* Proceedings of the International Symposium on Tropical Peatland, 6–10 May 1991, Kuching, Malaysian Agricultural Research and Development Institute, Kuala Lumpur, 7–16.

Naidu, G. (1997). The Straits of Malacca in the Malaysian economy. In: A. Hamzan (Ed.), *The Straits of Malacca: international co-operation in trade, funding and navigational safety* (pp. 33–60). Kuala Lumpur: Pelanduk Publications, Maritime Institute of Malaysia.

Nair, R. & Siew, N.F. (1996). Economy and sea-borne trade: global and national outlook and implications for the development of the Malaysia's merchant fleet. *MIMA Bulletin,* 3, 2: 22–25.

Neidpath, J. (1981). *The Singapore naval base and the defence of Britain's eastern empire, 1919–1941.* Oxford: Clarendon Press.

Newman, H. (1992). Operation 'Reef Rescue'. *Reef Encounter,* 10: 18–19.

New Straits Times (1988). Pesticide residues. 23 November.

New Straits Times (1988). Pesticide residue in the Cameron Highlands. 9 December.

New Straits Times (1997). Fishermen upset over trawler encroachment. 20 November.

New Straits Times (1998). 400 factories treating wastes at integrated toxic waste centre—Shah Alam—incinerator. 24 June.

New Straits Times (1998). Pasir Gudang factories—pollution. 29 June.

New Straits Times (1998). Toxic waste drums still lying around village sites. 29 December.

New Sunday Times (1998). Illegal trawling—nets, traps destroyed. 22 February.

New Sunday Times, Sunday Magazine (1997). Special Issue. Salient Straits. 7 December.

Nichol, J. (1993). Remote sensing of tropical blackwater rivers: a model for environmental water quality analysis. *Applied Geography*, 13: 153–168.

Nickerson, D., Hiew, K. & Chong, G. (1997). Special area management for conservation and sustained production of marine biodiversity in Malaysia. *Tropical Coasts*, 4, 1: 12–15.

Nik Fuad, N.A. (1988). Package treatment plant using RBC. Proceedings of the 14th WEDC Conference on Water and Urban Services in Asia and the Pacific, 11–15 April 1988, KL, WEDC Loughborough Uniiversity of Technology, England.

Ninkovich, D. (1976). Late Cenozoic clockwise rotation of Sumatra. *Earth Planetary Science Letters*, 29: 269–275.

Ninkovich, D., Shackleton, N.J., Abdel-Monem, A.A., Obradovich, J.D. & Izett, G. (1978). K-Ar age of the late Pleistocene eruption of Toba, north Sumatra. *Nature*, 276: 574–577.

Norhayati, M.T. (1997). Fate of spilled oil and ecological and socioeconomic impacts of oil pollution in the Straits of Malacca. In: H. Yu, K. Sung Low, N. Minh Son & D.-Y. Lee (Eds), *Oil spill modelling in the East Asian region: with special reference to the Straits of Malacca* (pp. 242–263). Quezon City: GEF/UNDP/IMO.

Norhayati M.T., Mohd Sufian, M.N., Mohammad Radzi, A., Zanariah, A. & Yeap, E.B. (1994). Monitoring of leachate generated from the municipal landfills in the Klang valley. Proceedings of the International Conference on Environment and Development, Kuala Lumpur, Malaysia, 689–695.

Nossin, J.J. (1961). Relief and coastal development in northwestern Johore. *Journal of Tropical Geography*, 15: 27–39.

OCIMF/ICS Publications (1990). *Straits of Malacca and Singapore—a guide to planned transits by deep draught vessels* (3rd edn). OCIMF/ICS Publications.

Ong, K.S. (Undated). *Aquaculture development in Malaysia in the 1990s.* Penang: Fisheries Research Institute.

Ong, K.S. (1985). Research as a support service for aquaculture development. Report on Aquaculture Conference, Department of Fisheries, Ministry of Agriculture, Kuala Lumpur, 9–16.

Ongkosongo, O.S.R. (1982). The nature of coastline changes in Indonesia. *The Indonesian Journal of Geography*, 12, 43: 1–22.

Ono, A. (1997). Japan's contribution to safety and pollution mitigation in the Straits of Malacca. In: A. Hamzah (Ed.), *The Straits of Malacca: international co-operation in trade, funding and navigational safety* (pp. 241–246). Kuala Lumpur: Pelanduk Publications, Maritime Institute of Malaysia.

Ooi, G.G. & Lo, N.P. (1990). Toxicity of herbicides to Malaysian rice field fish. Proceedings of the 3rd International Symposium on Plant Protection in the Tropics.

Ooi, H.S. (1992). Research in mechanisation on peat soil in Malaysia. In: B.Y. Aminuddin (Ed.), *Tropical peat.* Proceedings of the International Symposium on Tropical Peatland, 6–10 May 1991, Kuching, Malaysian Agricultural Research and Development Institute, Kuala Lumpur, 239–243.

Ooi, J.B. (1955). Tin mining landscape of Kinta: a study to some of the major environmental problems of mining in the Kinta Valley. *Journal of Tropical Geography*, 4: 58.

Osborne, M. (1988). *Southeast Asia.* Sydney: Allen and Unwin.

Otto, S.R., Ongkosongo, O.S.R. & Lukman, E. (1993). Environmental changes with a rising sea level on the coast of Batam Island. *MJTG Special Issue*, 24, 1/2: 99–102.

Packham, G.H. (1993). Plate tectonics and the development of sedimentary basins of the dextral regime in western Southeast Asia. *Journal of South East Asian Sciences*, 8, 1–4: 497–511.

Parry, J.H. (1973). *The Age of Reconnaisance.* London: Sphere.

Paton, J.R. (1964). The origin of limestone hills of Malaya. *JTG*, 18: 134–147

Peet, G. (1997). Financing Straits management: policy options. In: A. Hamzah (Ed.), *The Straits of Malacca: international co-operation in trade, funding and navigational safety* (pp. 145–164). Kuala Lumpur: Pelanduk Publications, Maritime Institute of Malaysia.

Peh, C.H. (1978). *Rates of sediment transport by surface wash in three forested areas of Peninsular Malaysia.* Occasional Paper 3. Kuala Lumpur: Department of Geography, University of Malaya.

Pelras, C. (1996). *The Bugis.* London: Routledge.

Pelzer K. (1968). Man's role in changing the landscape of Southeast Asia. *Journal of Asian Studies,* 27, 2: 269–279.

Pelzer, K. (1978). *Planter and peasant. Colonial policy and the agrarian struggle in East Sumatra.* S'Gravenhage: Martinus Nijhoff.

Perry, M. (1991). The Singapore growth triangle: state, capital and labour at a new frontier in the world economy. *Singapore Journal of Tropical Geography,* 12, 2: 138–151.

Perry, M. (1998). The Singapore growth triangle in the global and local economy. In: V. Savage, L. Kong & W. Neville (Eds), *The Naga awakens: growth and change in Southeast Asia* (pp. 87–110). Singapore: Times Academic Ltd.

Phillips, M.J., Kwei Lin, C. & Beveridge, M.C.M. (1993). Shrimp culture and the environment: lessons from the world's most rapidly expanding warmwater aquaculture sector. In: R.S.V. Pullin, H. Rosenthal & J.L. Maclean (Eds.) *Environment and aquaculture in developing countries.* ICLARM Conference Proceedings, 31: 171–197.

Port of Singapore Authority (1994). *Annual report.*

Powell, R. (1997). Erasing memory, inventing tradition, rewriting history: planning as a tool of ideology. In: B. Shaw & R. Jones (Eds), *Contested urban heritage—voices from the periphery* (pp. 85–100). Aldershot: Ashgate.

Pramojanee, P., Hasting, P., Liengsakul, M. & Engakul, V. (1986). The Holocene transgression in Peninsular Thailand. *Geosea V, Proceedings of the Geological Society of Malaysia,* 1: 551–564.

Prijosoesilo, Y., Sunarya & Wahab, A. (1993). Recent progress of geological investigations in Indonesia. *Journal of Southeast Asian Earth Sciences,* 8: 1: 25–36.

Proctor, J. (1983). Mineral nutrients in tropical forest. *Progress in Physical Geography,* 7: 422–431.

PUTL (1974). *Tidal swamp reclamation—the second Five Year Development Plan, 1974/75–1978/79.* Jakarta: Ministry of Public Works and Electric Power.

Radjagukguk, B. (1992). Utilization and management of peatlands in Indonesia for agriculture and forestry. In: B.Y. Aminuddin (Ed.), *Tropical peat.* Proceedings of the International Symposium on Tropical Peatland, 6–10 May 1991, Kuching, Malaysian Agricultural Research and Development Institute, Kuala Lumpur, 21–27.

Raja Malik, R.K. (1995). Safety of the Malacca Strait and other outstanding issues. *Malaysian Institute of Maritime Affairs, Issue Paper No. 2,* Kuala Lumpur.

Raja Malik, R.K. (1996). Navigational safety in the Straits of Malacca and Singapore. Paper presented at a Conference on Navigational Safety and Control of Pollution in the Straits of Malacca and Singapore: Modalities of International Co-operation, 2–3 September 1996, Singapore, Institute of Policy Studies and IMO.

Rajah, S.S., Chand, F. & Santokh Singh, D. (1977). The granitoids and mineralization of the Eastern Belt of Peninsular Malaysia. *Bulletin of the Geological Society of Malaysia,* 9: 209–232.

Reid, A. (1988). *Southeast Asia in the age of commerce 1450–1680: Volume 1: The lands below winds.* New Haven: Yale University Press.

Reid, A. (1993). *Southeast Asia in the age of commerce 1450–1680: Volume 2: Expansion and crisis.* New Haven: Yale University Press.

Reid, A. & Castles, L. (1975). Pre-colonial state systems in Southeast Asia. *Journal of the Malay Branch, Royal Asiatic Society,* Monograph 6.

Rice, R.C. (1989). Riau and Jambi: rapid growth in dualistic natural resource-intensive economies. In: H. Hill (Ed.), *Unity and diversity: regional economic development in Indonesia since 1970* (pp. 125–150). Singapore: Oxford University Press.

Rice, R.C. (1991). Environmental degradation, pollution, and the exploitation of Indonesia's fishery resources. In: J. Hardjono (Ed.), *Indonesia: resources, ecology, and environment.* Singapore: Oxford University Press.

Rieley, J. (1992). The ecology of tropical peatswamp forest—a southeast Asian perspective. In: B.Y. Aminuddin (Ed.), *Tropical peat.* Proceedings of the International Symposium on Tropical Peatland, 6–10 May 1991, Kuching, Malaysian Agricultural Research and Development Institute, Kuala Lumpur, 244–254.

Rigg, J. (1991). *Southeast Asia: a region in transition.* London: Unwin Hyman.

Rigg, J. (1997). *Southeast Asia: the human landscape of modernisation and development.* London: Routledge.

Robinson, R. (1997). Shipping and international trade in the Asian region: an international view. In: A. Hamzah (Ed.), *The Straits of Malacca: international co-operation in trade, funding and navigational safety* (pp. 263–282). Kuala Lumpur: Pelanduk Publications, Maritime Institute of Malaysia.

Rodan (1987). *The political economy of Singapore's growth.* London: Macmillan.

Ross, M.S. (1980). The role of land clearing in Indonesia's transmigration program. *Bulletin of Indonesian Economic Studies,* 16, 1: 75–85.

Sandhu, K.S. & Wheatley, P. (Eds) (1983). *Melaka: the transformation of a Malay capital, c. 1400–1800.* 2 vols., Kuala Lumpur: Oxford University Press.

SarDesai, D. (1994). *Southeast Asia past and present.* Boulder: Westview Press.

Savage, V.R., Kong, L. & Neville, W. (1998), *The Naga awakens: growth and change in Southeast Asia.* Singapore: Times Academic.

Schwamborn, R. & Saint-Paul, U. (1996). Mangroves—forgotten forests? *Natural Resources and Development,* 43, 44: 13–36.

Schwartz, M.O., Rajah, S.S., Askury, A.K., Putthapiban, P. & Djaswadi, S. (1995). The Southeast Asian tin belt. *Earth Science Reviews,* 38: 95–293.

Scrivenor, J.B. (1931). *The geology of Malaya.* London: MacMillan.

Scrivenor, J.B. & Wilbourne, E.S. (1923). The geology of Langkawi islands with a geological sketch map. *Journal of the Malay Branch, Royal Asiatic Society,* 1: 338–347.

Seow, C.W. (1985). Large scale intensive prawn farming. *Aquaculture Conference,* March, Johor Bahru.

Sewerage and Drainage Department (undated). Guidelines for planning and approval for sanitary installation, sewer, sewage treatment works and main drainage within the Federal Territory, Kuala Lumpur. Kuala Lumpur: City Hall.

Shallow, P.G.D. (1956). River flow in the Cameron Highlands. *Hydroelectric Technical Memoir 3,* Kuala Lumpur.

Sham, S. (1991). Urban planning for air quality management: a Malaysian perspective. *Akademika,* 39: 125–149.

Sham, S. (1997). *Environmental Quality Act 1974: then and now.* Bangi: Universiti Kebangsaan Malaysia.

Shaw, B. & Jones, R. (Eds) (1997). *Contested urban heritage—voices from the periphery.* Aldershot: Ashgate.

Siddayao, C.M. (1978). *The off-shore petroleum resources of Southeast Asia: potential conflict situations and related economic situations.* Kuala Lumpur: Oxford University Press.

Sidik, O. (1995). A vessel traffic system for the Straits of Malacca, Singapore and Johor. Paper presented at the Conference on VTS at Maritime Institute of Malaysia (MIMA), Kuala Lumpur.

Sinclair, K. (1967). Hobson and Lenin in Johore: Colonial Office policy towards British concessionaires and investors, 1878–1907. *Modern Asian Studies*, 1, 4: 335–352.

Sinsakul, S. (1992). Evidence of Quaternary sea level changes in the coastal areas of Thailand: a review. *Journal of Southeast Asian Earth Sciences*, 7: 23–37.

SIRIM (Standards and Industrial Research Institute Malaysia) (1989). Draft Malaysia Standard Code of Practice for Design and Installation of Sewerage Systems.

Smith, R.W. & Roach, J.A. (1996). *Navigation rights and responsibilities in international straits: a focus on the Straits of Malacca.* Malaysia: Malaysian Institute of Maritime Affairs, Issue Paper.

Soegiarto, A. (1987). Marine and coastal environment problems in Indonesia: an input to the assessment of East Asian regional seas.

Soegiarto, A. (1993). ASEAN cooperation in marine science: review of programmes, results and achievements. *ASEAN Journal of Science Technology and Development*, 10, 1: 1–13.

Soekardi, M. & Hidayat, A. (1988). *Extent and distribution of peat soils of Indonesia.* Bogor: Third Meeting of Cooperative Research on Problem Soils, CRIFC.

Soemartowo, O. (1991). Human ecology in Indonesia: the search for sustainability in development. In: J. Harjono (Ed.), *Indonesia: resources, ecology and environment.* Singapore: Oxford University Press.

Sopher, D. (1987). *The sea nomads.* Singapore: National Museum.

Suki, A. (1993). An overview of sewage problems, management and control technologies. In: B.G. Yeoh *et al* (Eds) *Waste management in Malaysia: current status and prospects for bioremediation* (pp. 173–193). Malaysia: Ministry of Science, Technology and the Environment.

Suki, A. & Awang, M. (1990). Oxidation pond and aerated lagoon. Paper presented at Seminar on Options for Sewage Treatment, 18 September 1990, ENSEARCH.

Suki, A., Jenny, H. & Yaziz, M.I. (1987). Performance of selected oxidation ponds. In: Malaya a field survey. Presented at the Seminar on Domestic Wastewater Treatment Alternatives, 9–11 November, Université Pertanian Malaysia, Kuala Lumpur.

Suki, A., Jenny, H. & Yaziz, M.I. (1989). Performance of selected oxidation ponds. In: A. Suki, M. Rashid Zainal & M. Awang (Eds), *Characteristics of a low load oxidation pond.* Proceedings of the 5th Chemical Symposium, Kuala Lumpur.

Suki, A. *et al* (1990). Mass balance visit of the oxidation ditch treatment plant (Taman Seri Gombak), 1989. Technical Advisory Service Report 01/90.

Sun, J. & Pan, T.-C. (1995). Seismic characteristics of Sumatra and its relevance to Peninsular Malaysia and Singapore. *Journal of Southeast Asian Earth Sciences*, 12, 1/2: 105–111.

Sutikno & Sunarto, K. (1992). Wetlands and their prospects in Indonesia. *Indonesian Journal of Geography*, 23–25, 64–66: 47–58.

Tam, T., Yeow, K., Poon, Y. (1982). *Land application of POME.* Proceedings of a Regional Workshop on Palm Oil Mill Technology and Effluent Treatment, Kuala Lumpur, 216–224.

Tan, C.K. (1985). Aquaculture extension service. Paper presented at Aquaculture Conference, 9–12 December 1985, Johor Bahru, Malaysia.

Tan, C.K. (1992). Aquaculture in Malaysia. Country report. Proceedings of the 1990 APO Symposium on Aquaculture in Asia, Taiwan Fisheries Research Institute, 105–110.

Tan, E.G. (1992). Status and trends of waste management in Singapore. In: T.-E. Chua & L.R. Garces (Eds), *Waste management in the coastal waters of the ASEAN region: roles of governments, banking institutions, donor agencies, private sector and communities.* ICLARM Conference Proceedings 33, Ministry of the Environment, and Canada-ASEAN Center Singapore; Asian Development Bank, and International Center for Living Aquatic Resources Management, Philippines.

Tan, P.K. (1992). Tourism in Penang: its impact and implications. In: P.K. Voon & Tunku Shamsul Bahrin (Eds), *The view from within* (pp. 263–278). Kuala Lumpur: Department of Geography, University of Malaya.

Tang, S.M., Orlic, I., Makjanic, J., Wu, X.K., Ng, T.H., Wong, M.K., Lee K.K. & Chen, N. (1996). *A survey of levels of metallic and organic pollutants in Singapore coastal waters and marine sediments.* ASEAN marine environmental management: quality criteria and monitoring for aquatic life and human health protection. ASEAN-Canada, CPMS-II End of Project Conference, 24–28 June 1996, Penang, Malaysia.

Tarling, N. (1962). *Anglo-Dutch rivalry in the Malay world 1780–1824.* St Lucia: University of Queensland Press.

Tarling, N. (1993). *The fall of imperial Britain in South-East Asia.* Singapore: Oxford University Press.

Tate, D.J. (1971). *The making of modern South-East Asia. Volume 1. The European conquest.* Kuala Lumpur: Oxford University Press.

Tate, D.J. (1979). *The making of modern South-East Asia. Volume 2. The western impact.* Kuala Lumpur: Oxford University Press.

Taylor, D.C. (1981). *The economics of Malaysian paddy production and irrigation.* Bangkok: The Agricultural Development Council.

Teh, T.S. (1989). *The permatang system in Peninsular Malaysia: an overview.* IGCP Project 274 International Symposium, Ipoh, 4–10 September 1989, 36–68.

Teh, T.S. (1993). Potential impacts of sea-level rise on the permatang coast of Peninsular Malaysia. *MJTG Special Issue,* 24, 1/2: 41–56.

Teh T.S. & Lim, C.H. (1993). Impacts of sea level rise on the mangroves of Peninsular Malaysia. *Malaysian Journal of Tropical Geography,* Special Issue, 24, 1: 57–72.

Tengku Ubaidullah, A.K. (1985). Aquaculture Development Strategy and Programme. Report on the Aquaculture Conference, Fisheries Department, Ministry of Agriculture, Malaysia, 4–8.

The Star (1993). Research from UTM—high levels of meavy metals in shellfish. 17 May.

The Straits Times (1997). Turtle excluder device. 28 February.

Thoburn, J. (1978). Malaysia's tin supply problems. *Resources Policy,* 4: 1: 31–38.

Tie, Y.L. & Lim, J.S. (1992). Characteristics and classification of organic soils on Malaysia. In: B.Y. Aminuddin (Ed.), *Tropical peat.* Proceeding of the International Symposium on Tropical Peatland, 6–10 May, 1991, Kuching, Malaysian Agricultural Research and Development Institute, 107–113.

Tjia, H.D. (1970). Quaternary shorelines of the Sunda Land. *Southeast Asia. Geol. en Mijn,* 49: 135–144.

Tjia, H.D. (1973). Geomorphology. In: D.J. Gobbett & C.S. Hutchison (Eds), *Geology of the Malay Peninsula* (pp. 135–144). Chichester: Wiley-Interscience.

Tjia, H.D. & Fujii, S. (1989). The coastal zone of Peninsular Malaysia. IGCP Project 724 International Symposium, Ipoh, 4–10 September 1989, 69–84.

Tjia, H.D., Fujii, S., Kigoshi, K. & Sugimura, A. (1975). Additional dates on raised shorelines in Malaysia and Indonesia. *Sains Malaysiana*, 4: 69–84.

Tjia, H.D., Fujii, S., Kigoshi, K., Sugimura, A. & Zakaria, T. (1972). Radiocarbon dates of elevated shorelines, Indonesia and Malaysia. *Part I, Quaternary Research*, 2: 487–495.

Tomascik, T., Mah, A.J., Nontji, A. & Moosa, M.K. (Eds) (1997). *The ecology of the Indonesia seas, Part One and Part Two.* Halifax, Nova Scotia: Dalhousie University, Periplus Editions (HK) Ltd.

Tregonning, K.G. (1967). *Home Port Singapore: a history of the Straits Steamship Company Limited, 1890–1965.* Singapore: Oxford University Press.

Tregonning, K.G. (undated). *Straits tin. A brief account of the first seventy-five years of the Straits Trading Company Ltd 1887–1962.* Singapore: Straits Times Press.

Trocki, C. (1979). *Prince of pirates. The Temmenggongs and the development of Johor and Singapore, 1784–1885.* Singapore: Singapore University Press.

Tsunehike, Y. & Keiichiro Nakagawa (Eds) (1985). *Business history of shipping: strategy and structure.* Tokyo: University of Tokyo Press.

Tunku Shamsul, B. & Teh, T.S. (1993). Coastal land reclamation as a counter-attack policy option in response to sea level rise in Malaysia. *MJTG Special Issue*, 24: 1/2: 73–82.

Turnbull, C.M. (1989). *A History of Singapore, 1879–1988.* Singapore: Oxford University Press.

Ungar, R. (1996). Ships of the Middle Ages. In: J. Hattendorf (Ed.), *Maritime history: 1: The age of discovery* (pp. 35–49). Malabar, Florida: Krieger.

United Nations Environmental Programme (1982). *GESAMP: The health of the oceans.* UNEP Regional Seas Reports and Studies No. 16, Geneva.

Valencia, M.J. (1985). *Southeast Asian seas—oil under troubled waters: hydrocarbon potentials, jurisdictional issues and international relations.* Singapore: Oxford University Press.

Valencia, M.J. (1991). Coastal area management in ASEAN: The Trans-national issue. In: L.M. Chou *et al* (Eds.), *Towards an integrated management of tropical coastal resources.* ICLARM Conference Proceedings 22, National University of Singapore, 279–290.

Van Dyke, J. (1997). Legal and political problems governing international straits. In: A. Hamzah (Ed.), *The Straits of Malacca: international co-operation in trade, funding and navigational safety* (pp. 305–326). Pelanduk Publications, Maritime Institute of Malaysia.

Verstappen, H.T. (1964). Geomorphology of Sumatra. *Journal of Tropical Geography*, 18: 184–191.

Verstappen, H.T. (1973). *A geomorphological reconnaissance of Sumatra and adjacent islands.* Groningen: Wolters-Noordhoff.

Verstappen, H.T. (1980) Quaternary climatic changes and the natural environment in SE Asia. *Geojournal*, 4: 45–54.

Voon, P.K. (1976). *Western rubber planting enterprise in Southeast Asia, 1876–1921.* Kuala Lumpur: Penerbit University Malaya.

Wan, S. & Sim Lin Woon, (1996). *Safety of navigation in the Straits of Malacca and Singapore.* Singapore: Institute of Policy Studies.

Wang Gungwu (1964). The opening of relations between China and Malacca, 1403–5. In: J. Bastin & R. Roolvink (Eds), *Malayan and Indonesian studies* (pp. 87–104). Oxford: Clarendon Press.

Wang, H. (1992). Isostasy and Holocene high sea levels in East and Southeast Asia. *Journal of Southeast Asian Earth Sciences*, 7, 1: 17–22.

Watts, I.E.M. (1954). Line-squalls of Malaya. In: J.B. Ooi & L.S. Chia (Eds), *Readings on the climate of West Malaysia and Singapore* (pp. 3–15). Singapore: Oxford University Press.

Wee, Y.C. (1982). Mangrove ecosystem in Singapore. In: *Mangrove forest ecosystem productivity in Southeast Asia.* Biotrop Special Publication No. 17, 93–97.

Welch, D.N. & Mohd Adnan, M.N. (1989). Drainage works on peat in relation to crop cultivation—a review of problems. Proceedings of the National Seminar on Soil Management for Food and Fruit Crop Production, March 1989, Kuala Lumpur, 96–110.

Wheatley, P. (1961). *The golden Khersonese.* Kuala Lumpur: Oxford University Press.

Whitmore, T.C. (1984). *Tropical rain forests of the Far East* (2nd edn). Oxford: Clarendon Press.

Whitmore, T.C. (1990). *An introduction to tropical rain forests.* Oxford: Clarendon Press.

Widhyawan Prawiraatmadja (1997). Indonesia's transition to a net oil importing country: critical issues in the downstream oil sector. *Bulletin of Indonesian Economic Studies,* 33, 2: 49–71.

Withington, W.A. (1985). The intermediate city concept reviewed and applied to major cities in Sumatra, Indonesia. *Indonesian Journal of Geography,* 15–16, 49–51: 1–14.

Wolters (1967). *Early Indonesian commerce.* Ithaca: Cornell University Press.

Wong, J. (1996). The marine leisure industry: best environmental practices. *MIMA Bulletin,* 3: 2: 14–21.

Wong Lin Ken (1960). The Trade of Singapore 1819–1869. *JMBRAS,* 33, 4: 5–308.

Wong, P.P. (1990). Impact of sea level rise on the coasts of Singapore: preliminary observations. *Journal of Southeast Asian Earth Sciences,* 7, 1: 65–70.

Wong, P.P. (1993). Island tourism development in Peninsular Malaysia: environmental perspective. In: P.P. Wong (Ed.), *Tourism vs environment: the case for coastal areas* (pp. 83–97). Borfrecht, Boston, London: Kluwer Academic Publications.

Wong T.C. & Goh, K.C. (1996). The tin market collapse in the Kinta Valley, Malaysia—an impact assessment a decade after. *Asia-Pacific Viewpoint,* 37, 1: 81–88.

Woodroffe, C.D. (1992). Mangrove sediments and geomorphology. In: A.I. Robertson & D.M. Alongi (Eds), *Tropical mangrove systems. Coastal and estuarine studies,* 41 (pp. 7–42). Washington: American Geophysical Union.

Woodroffe, C.D. (1993). Late Quaternary evolution of coastal and lowland riverine plains of Southeast Asia and northern Australia: an overview. *Sedimentary Geology,* 83: 163–175.

World Bank (1994). *Indonesia—sustaining development: a World Bank country study.* Washington.

World Resources Institute (1996). *World resources 1996–1997.* Oxford: Oxford University Press.

Wosten, J.H.M., Ismail, A.B. & van Wijk, A.L.M. (1997). Peat subsidence and its practical implications: a case study in Malaysia. *GEODERMA,* 78: 25–36.

Yancey, T.E. (1973). Holocene radiocarbon dates on the 3-metre wave cut notch in north western Peninsular Malaysia. *Geological Society of Malaysia Newsletter,* 45: 8–11.

Yap, L. (1977). A socio-economic analysis of the problems of over expansion on the west coast of Peninsular Malaysia. In: *Malaysian fisheries—a diminishing resource* (pp. 29–41). Penang: Consumers' Association of Penang.

Yaziz, M.I. (1981). Occurrence of antibiotic resistant *salmonella* in sewage and the effect of primary sedimentation on their number. *Pertanika,* 4: 1: 39–42

Yeap, E.B. (1993). Tin and gold mineralisations in Peninsular Malaysia and their relationships to the tectonic development. *Journal of Southeast Asian Earth Sciences,* 8, 1–4: 329–348.

Yeap, E.B., Tan, B.K. & Chow, W.S. (1993). Geotechnical aspects of development over reclaimed former alluvial mining land and ponds in Malaysia. *Journal of Southeast Asian Earth Sciences,* 8: 1–4.

Yeoh, B.G. (1996). *Contesting space: power relations and urban built environment in colonial Singapore.* Kuala Lumpur: Oxford University Press.

Yeoh, B.G. & Kong, L. (Eds) (1995). *Portraits of places—history, community and identity in Singapore.* Singapore: Times Editions.

Yip, W.K., Loo, M., Hsu, L., Chou, L.M. & Khoo, H.W. (1987). Conditions and life in the Singapore River. *Singapore Scientist,* 13, 3: 59–64.

Zaid, I. (1993). Wastes from rubber processing and rubber product manufacturing industries. In: B.G. Yeoh *et al* (Eds), *Waste management in Malaysia: current status and prospects for bioremediation* (pp. 137–151). Malaysia: Ministry of Science, Technology and the Environment.

Zauyah, S. (1986). Characterisation of some weathering profiles on metamorphic rocks in Peninsular Malaysia. PhD Thesis, University of Ghent.

INDEX

For Product Safety Concerns and Information please contact our EU representative GPSR@taylorandfrancis.com Taylor & Francis Verlag GmbH, Kaufingerstraße 24, 80331 München, Germany

Printed and bound by CPI Group (UK) Ltd, Croydon, CR0 4YY
01/05/2025
01858365-0001